Quantum Computing
from the Ground Up

Quantum Computing from the Ground Up

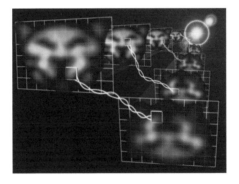

Riley Tipton Perry

University of New South Wales, Australia

World Scientific

NEW JERSEY · LONDON · SINGAPORE · BEIJING · SHANGHAI · HONG KONG · TAIPEI · CHENNAI

Published by

World Scientific Publishing Co. Pte. Ltd.

5 Toh Tuck Link, Singapore 596224

USA office: 27 Warren Street, Suite 401-402, Hackensack, NJ 07601

UK office: 57 Shelton Street, Covent Garden, London WC2H 9HE

British Library Cataloguing-in-Publication Data
A catalogue record for this book is available from the British Library.

ISBN 978-981-4412-11-7 (pbk)

Printed in Singapore by Mainland Press Pte Ltd.

Contents

Acknowledgments   xi

1. Introduction   1

   1.1   What is Quantum Computing?   1

   1.2   Why Another Quantum Computing Tutorial?   1

       1.2.1   Quantum Computation and Quantum Information   2

2. Computer Science   3

   2.1   Introduction   3

   2.2   History   3

   2.3   Turing Machines   5

       2.3.1   Binary Numbers and Formal Languages   7

       2.3.2   Turing Machines in Action   9

       2.3.3   The Universal Turing Machine   10

       2.3.4   The Halting Problem   11

   2.4   Circuits   13

       2.4.1   Common Gates   13

       2.4.2   Combinations of Gates   15

       2.4.3   Relevant Properties   16

       2.4.4   Universality   16

   2.5   Computational Resources and Efficiency   17

       2.5.1   Quantifying Computational Resources   19

       2.5.2   Standard Complexity Classes   20

       2.5.3   The Strong Church–Turing Thesis   22

       2.5.4   Quantum Turing Machines   23

   2.6   Energy and Computation   24

2.6.1	Reversibility	24
2.6.2	Irreversibility	24
2.6.3	Landauer's Principle	24
2.6.4	Maxwell's Demon	25
2.6.5	Reversible Computation	26
2.6.6	Reversible Gates	26
2.6.7	Reversible Circuits	29

3. Mathematics for Quantum Computing   31

3.1	Introduction	31
3.2	Polynomials	32
3.3	Logical Symbols	32
3.4	Trigonometry Review	33
	3.4.1 Right Angled Triangles	33
	3.4.2 Converting Between Degrees and Radians	33
	3.4.3 Inverses	34
	3.4.4 Angles in Other Quadrants	34
	3.4.5 Visualisations and Identities	35
3.5	Logs	37
3.6	Complex Numbers	37
	3.6.1 Polar Coordinates and Complex Conjugates	39
	3.6.2 Rationalising and Dividing	43
	3.6.3 Exponential Form	43
3.7	Matrices	45
	3.7.1 Matrix Operations	46
3.8	Vectors and Vector Spaces	50
	3.8.1 Introduction	50
	3.8.2 Column Notation	53
	3.8.3 The Zero Vector	54
	3.8.4 Properties of Vectors in $\mathbb{C}^n$	54
	3.8.5 The Dual Vector	55
	3.8.6 Linear Combinations	56
	3.8.7 Linear Independence	57
	3.8.8 Spanning Set	57
	3.8.9 Basis	57
	3.8.10 Probability Theory	58
	3.8.11 Probability Amplitudes	59
	3.8.12 The Inner Product	60
	3.8.13 Orthogonality	63

3.8.14 The Unit Vector 64

3.8.15 Bases for $\mathbb{C}^n$ 65

3.8.16 The Gram Schmidt Method 67

3.8.17 Linear Operators 67

3.8.18 Outer Products and Projectors 68

3.8.19 The Adjoint 72

3.8.20 Eigenvalues and Eigenvectors 74

3.8.21 Trace 75

3.8.22 Normal Operators 77

3.8.23 Unitary Operators 78

3.8.24 Hermitian and Positive Operators 80

3.8.25 Diagonalisable Matrix 80

3.8.26 The Commutator and Anti-Commutator 81

3.8.27 Polar Decomposition 82

3.8.28 Spectral Decomposition 82

3.8.29 Tensor Products 83

3.9 Fourier Transforms 85

3.9.1 The Fourier Series 86

3.9.2 The Discrete Fourier Transform 89

4. Quantum Mechanics 93

4.1 History 94

4.1.1 Classical Physics 94

4.1.2 Important Concepts 95

4.1.3 Statistical Mechanics 97

4.1.4 Important Experiments 98

4.1.5 The Photoelectric Effect 100

4.1.6 Bright Line Spectra 101

4.1.7 Proto Quantum Mechanics 102

4.1.8 The New Theory of Quantum Mechanics 105

4.2 Important Principles for Quantum Computing 109

4.2.1 Linear Algebra 110

4.2.2 Superposition 110

4.2.3 Dirac Notation 111

4.2.4 Representing Information 112

4.2.5 Uncertainty 113

4.2.6 Entanglement 113

5. Quantum Computing   115

    5.1 Elements of Quantum Computing   115
         5.1.1 Introduction 115
         5.1.2 History 115
         5.1.3 Bits and Qubits 116
         5.1.4 Entangled States 131
         5.1.5 Quantum Circuits 133
    5.2 Important Properties of Quantum Circuits   147
         5.2.1 Common Circuits   148
    5.3 The Reality of Building Circuits 154
         5.3.1 Building a Programmable Quantum Computer . .   154
    5.4 The Four Postulates of Quantum Mechanics   155
         5.4.1 Postulate One 155
         5.4.2 Postulate Two   156
         5.4.3 Postulate Three   157
         5.4.4 Postulate Four 160

6. Information Theory   163

    6.1 Introduction   163
    6.2 History   164
    6.3 Shannon's Communication Model   164
         6.3.1 Channel Capacity   165
    6.4 Classical Information Sources 166
         6.4.1 Independent Information Sources   166
    6.5 Classical Redundancy and Compression   168
         6.5.1 Shannon's Noiseless Coding Theorem   169
         6.5.2 Quantum Information Sources   171
         6.5.3 Pure and Mixed States   171
         6.5.4 Schumacher's Quantum Noiseless Coding Theorem   172
    6.6 Noise and Error Correction   179
         6.6.1 Quantum Noise   181
         6.6.2 Quantum Error Correction   181
    6.7 Bell States 188
         6.7.1 Same Measurement Direction   189
         6.7.2 Different Measurement Directions 190
         6.7.3 Bell's Inequality 191
    6.8 Cryptology 195
         6.8.1 Classical Cryptography 195

        6.8.2   Quantum Cryptography   196
    6.9   Alternative Models of Computation   200

7. Quantum Algorithms   201

        7.0.1   Introduction   201
    7.1   Deutsch's Algorithm   202
        7.1.1   The Problem Defined   202
        7.1.2   The Classical Solution   202
        7.1.3   The Quantum Solution   203
        7.1.4   Physical Implementations   207
    7.2   The Deutsch–Josza Algorithm   207
        7.2.1   The Problem Defined   207
        7.2.2   The Quantum Solution   208
    7.3   Shor's Algorithm   210
        7.3.1   The Quantum Fourier Transform   210
        7.3.2   Fast Factorisation   214
        7.3.3   Order Finding   215
    7.4   Grover's Algorithm   220
        7.4.1   The Travelling Salesman Problem   221
        7.4.2   Quantum Searching   221

8. Using Quantum Mechanical Devices and Recent Developments   225

    8.1   Introduction   225
    8.2   Physical Realisation   225
        8.2.1   Implementation Technologies   227
    8.3   Quantum Computer Languages   228
    8.4   Encryption Devices   230
    8.5   Recent Developments   230
        8.5.1   Hardware and Architecture   230
        8.5.2   Cryptography   231
        8.5.3   Algorithms   231

Bibliography   233

Index   239

Acknowledgments

This tutorial began life as something of an open book and has had many contributors and proof readers. I would like to thank everyone who has contributed. Special thanks go to the following people:

Waranyoo Pulsawat, my dear wife. Waranyoo has been a constant source of support, companionship, and inspiration.

Brian Lederer, who is directly responsible for some parts of this text. He wrote the section on Bell states, some parts of the chapter on quantum mechanics, and a substantial amount of the other chapters. Without his help this work would never have been completed.

Mohamed Barakat and *Massoud Ghias Beygi* who have published their own version of the original book in Arabic!

Andreas Gunnarsson. Andreas' attention to detail is astonishing.

A special thanks to *Xerxes Rånby*. Xerxes had some valuable comments and found a number of errors.

All the members of the QC4Dummies Yahoo group (`http://groups.yahoo.com/group/QC4dummies/`) and administrators *David Morris* and *David Rickman*.

Sean Kaye and *Micheal Nielson* for mentioning this tutorial in their blogs.

The people at *Slashdot* (`http://slashdot.org/`) and *QubitNews* (`http://quantum.fis.ucm.es/`) for posting the tutorial for review.

James Hari, Simon Johnson, James Hollis, Nick Oosterhof, Rad Radish, Karol Bartkiewicz, Robin Kothari, Varun Vaidya, Kennedy Roulston, and Slashdotters *AC, Birdie 1013,* and *s/nemesis*.

Chapter 1

Introduction

1.1 What is Quantum Computing?

In quantum computers we exploit quantum effects to compute in ways that
are faster or more efficient than, or even impossible, on conventional computers. Quantum computers use a specific physical implementation to gain
a computational advantage over conventional computers. Properties called
superposition and entanglement may, in some cases, allow an exponential
amount of parallelism. Also, special purpose machines like quantum cryptographic devices use entanglement and other peculiarities like quantum
uncertainty.

Quantum computing combines quantum mechanics, information theory,
and aspects of computer science [31]. The field is a relatively new one that
promises secure data transfer, dramatic computing speed increases, and
may take component miniaturisation to its fundamental limit.

This text describes some of the introductory aspects of quantum computing. We'll examine some basic quantum mechanics, elementary quantum computing topics like qubits, quantum algorithms, physical realisations
of those algorithms, basic concepts from computer science (like complexity theory, Turing machines, and linear algebra), information theory, and
more.

1.2 Why Another Quantum Computing Tutorial?

Most of the books or papers on quantum computing require (or assume)
prior knowledge of certain areas like linear algebra or physics so the majority
of the current literature is hard to understand for the average computer
enthusiast or interested layman. This text attempts to teach basic quantum

computing from the ground up in an easily readable way. It contains a lot of the background in mathematics, physics, and computer science that you will need, although it is assumed that you know a little about computer programming.

At certain places in this document, topics that could make interesting research topics have been identified. These topics are presented in the following format:

Question   *The topic is presented in bold-italics.*

1.2.1   *Quantum Computation and Quantum Information*

By far the most complete book available for quantum computing is *Quantum Computation and Quantum Information* by Michael A. Nielsen and Isaac L. Chuang, which we'll abbreviate to QCQI. The main references for this work are QCQI and a great set of lecture notes, also written by Nielsen. Nielsen's lecture notes are currently available at `http://www.qinfo.org/people/nielsen/qicss.html`. An honourable mention goes out to Vadim V. Bulitko who has managed to condense a large part of QCQI into fourteen pages!

QCQI may be a little hard to get into at first, particularly for those without a strong background in mathematics. So this tutorial is, in part, a collection of worked examples from various web sites, sets of lecture notes, journal entries, papers, and books which may aid in understanding of some of the concepts in QCQI.

Chapter 2

Computer Science

2.1 Introduction

The special properties of quantum computers force us to rethink some of the most fundamental aspects of computer science. In this chapter we'll see how quantum effects can give us a new kind of Turing machine, new kinds of circuits, and new kinds of complexity classes. This is important as it was thought that these things were not affected by what a computer is built from, but it turns out that they are.

A distinction has been made between computer science and information theory. Although information theory can be seen as a part of computer science it is treated separately in this text with its own dedicated chapter. This is because the quantum aspects of information theory require some of the concepts introduced in the chapters that follow this one.

There's also a little mathematics and notation used in this chapter which is presented in the first few sections of chapter 3 and some basic C and Javascript code for which you may need an external reference.

2.2 History

The origins of computer science can be traced back to the invention of algorithms like Euclid's Algorithm (c. 300 BC), which is an algorithm for finding the greatest common divisor of two numbers. There are also much older sources like early Babylonian cuneiform tablets (c. 2000–1700 BC) that contain clear evidence of algorithmic processes [22]. Up until the 19th century it's difficult to separate computer science from other sciences like mathematics and engineering so we'll say here that computer science began as a separate science in the 19th century.

Fig. 2.1   Charles Babbage and Ada Byron.

An important precursor to early computing machines was the punched card. The Jacquard loom, which was invented by Joseph Marie Jacquard (1752–1834) in 1801 [35] made use of complex patterns stored on punched cards to control a sequence of operations. In the mid 19th century Charles Babbage, 1791–1871 (figure 2.1) designed and partially built several programmable computing machines (see figure 2.4 for the difference engine built in 1822). These machines had many of the features of modern computers. One of these machines called the analytical engine had removable programs on punch cards based on those used in the Jacquard loom. Babbage's friend, Ada Augusta King, Countess of Lovelace, 1815–1852 (figure 2.1), the daughter of Lord Byron, is considered by some as the first programmer for her writings on the Analytical engine. Sadly Babbage's work was largely forgotten until the 1930s and the advent of modern computer science. Modern computer science can be said to have started in 1936 when logician Alan Turing, 1912–1954 (figure 2.2) wrote a paper which contained the notion of a *universal computer*.

Fig. 2.2   Alan Turing and Alonzo Church.

The first electronic computers were developed in the 1940s and led Jon Von Neumann, 1903–1957 (figure 2.3) to develop a generic architecture on which many modern computers are based. *Von Neumann architecture* specifies an Arithmetic Logic Unit (ALU), control unit, memory, input/output (IO), a bus, and a computing process. The architecture originated in 1945 in the first draft of a report on EDVAC [10].

Fig. 2.3 Jon Von Neumann.

Computers increased in power and versatility rapidly over the next sixty years, partly due to the development of the transistor in 1947, integrated circuits in 1959, and increasingly intuitive user interfaces. Gordon Moore proposed Moore's law in 1965, the current version of which states that processor complexity will double every eighteen months with respect to cost (in reality it's more like two years). This law still holds but is starting to falter, as components are getting smaller. Soon they will be so small, only being made up of a few atoms [4], that quantum effects will become unavoidable, possibly ending Moore's law.

There are ways in which we can use quantum effects to our advantage in a classical sense, but by fully utilising those effects we can achieve much more. This approach is the basis for quantum computing.

2.3 Turing Machines

In 1928 David Hilbert, 1862–1943 (figure 2.5) asked if there was a universal algorithmic process to decide whether any mathematical proposition was true. His intuition suggested yes, then, in 1930 he went as far as claiming that there were no unsolvable problems in mathematics. This was promptly refuted by Kurt Gödel, 1908–1976 (figure 2.5) in 1931 by way of his *incompleteness theorem* which can be roughly summed up as follows:

Fig. 2.4   Babbage's difference engine.

> You might be able to prove every conceivable statement about
> numbers within a system by going outside the system in order
> to come up with new rules and axioms, but by doing so you'll
> only create a larger system with its own unprovable statements
> [24].

Then, in 1936 Alan Turing and Alonzo Church, 1903–1995 (figure 2.2) independently came up with models of computation, aimed at resolving whether or not mathematics contained problems that were *uncomputable*. These were problems for which there were no algorithmic solutions. An algorithm (which is a procedure for solving a mathematical problem) is guaranteed to end after a number of steps. Turing's model, now called a called a *Turing Machine* (TM), is depicted in figure 2.6. It turned out that the models of Turing and Church were equivalent in power. The thesis that any algorithm capable of being devised can be run on a Turing machine, as Turing's model was subsequently called, was given the names of both these pioneers, the *Church–Turing thesis* [30].

Fig. 2.5   David Hilbert and Kurt Gödel.

2.3.1   *Binary Numbers and Formal Languages*

Before defining a Turing machine we need to say something about *binary numbers*, since this is usually (although not confined to) the format in which data is presented to a Turing machine (see the tape in figure 2.6).

2.3.1.1   *Binary Representation*

Computers represent numbers in binary form, as a series of zeros and ones, because this is easy to implement in hardware compared with other forms, e.g. decimal. Any information can be converted to and from zeros and ones. We call this representation a *binary representation*.

Example   Here are some binary numbers and their decimal equivalents:

The binary number 1110 in decimal is 14.
Decimal number 212 when converted to binary becomes 11010100.

The binary numbers (on the left hand side) that represent the decimals 0–4 are as follows:

$$0 = 0$$
$$1 = 1$$
$$10 = 2$$
$$11 = 3$$
$$100 = 4$$

A binary number has the form $b_{n-1} \ldots b_2 b_1 b_0$ where n is the number of binary digits (or *bits*, with each digit being a 0 or a 1) and b_0 is the *least significant digit*. We can convert the binary string to a decimal number D using the following formula:

$$D = 2^{n-1}(b_{n-1}) + \ldots + 2^2(b_2) + 2^1(b_1) + 2^0(b_0). \tag{2.1}$$

Here is another example:

Example Converting the binary number 11010100 to decimal:

$$D = 2^7(1) + 2^6(1) + 2^5(0) + 2^4(1) + 2^3(0) + 2^2(1) + 2^1(0) + 2^0(0)$$
$$= 128 + 64 + 16 + 4$$
$$= 212$$

We call the binary numbers a base 2 number system because they are based on just two symbols, 0 and 1. By contrast, in decimal (which is base 10), we have $0, 1, 2, 3, \ldots, 9$.

All data in modern computers is stored in binary format; even machine instructions are in binary format. This allows both data and instructions to be stored in computer memory and it allows all of the fundamental logical operations of the machine to be represented as binary operations.

2.3.1.2 *Formal Languages*

Turing machines and other computer science models of computation use *formal languages* to represent their inputs and outputs. We say a language L has an alphabet $\sum$. The language is a subset of the set $\sum^*$ of all finite strings of symbols from $\sum$.

Example If $\sum = \{0, 1\}$ then the set of all even binary numbers $\{0, 10, 100, 110, \ldots\}$ is a langauge over $\sum$.

It turns out that the power of a computational model (or *automaton*) i.e. the class of algorithm that the model can implement, can be determined by considering a related question:

   What type of language can the automaton recognise?

A formal language in this setting is just a set of binary strings. In simple languages the strings all follow an obvious pattern, e.g. with the language:

$$\{01, 001, 0001, \ldots\}$$

the pattern is that we have one or more zeroes followed by a 1. If an automaton, when presented with any string from the language can read each symbol and then halt after the last symbol we say it recognises the language (providing it doesn't do this for strings not in the language). Then the power of the automaton is gauged by the complexity of the patterns for the languages it can recognise.

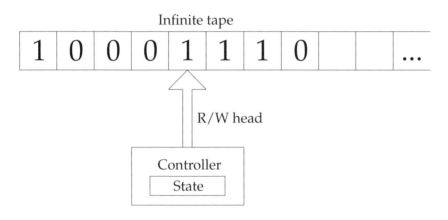

Fig. 2.6   A Turing machine.

2.3.2   *Turing Machines in Action*

A Turing Machine (which we'll sometimes abbreviate to TM) has the following components [12]:

(1) *A tape* — made up of cells containing $0, 1$, or blank. Note that this gives us a alphabet of $\sum = \{0, 1, \text{blank}\}$.
(2) *A read/write head* — reads or overwrites the current symbol on each step and moves one square to the left or right.
(3) *A controller* — controls the elements of the machine to do the following:
   (i) Read the current symbol.

(ii) Write a symbol by overwriting what's already there.

(iii) Move the tape left or right one square.

(iv) Change state.

(v) Halt.

4. *The controller's behaviour* — the way the TM switches states depending on the symbol it's reading, represented by a *Finite State Automata* (FSA).

The operation of a TM is best described by a simple example:

Example   Inversion inverts each input bit, for example:

$$001 \rightarrow 110$$

The behaviour of the machine can be represented by a two state FSA. The FSA is represented below in table form (where the states are labelled 1 and 2: 1 for the running state, and 2 for the halt state).

State	Value	New State	New Value	Direction
1	0	1	1	Move Right
1	1	1	0	Move Right
1	blank	2 - HALT	blank	Move Right
2 - HALT	N/A	N/A	N/A	N/A

2.3.3   *The Universal Turing Machine*

A *Universal Turning Machine* (UTM) is a TM (with an inbuilt mechanism described by a FSA) that is capable of reading, from a tape, a program describing the behaviour of another TM. The UTM simulates the ordinary TM, performing the behaviour generated when the ordinary TM acts upon its data. When the UTM halts its tape contains the result that would have been produced by the ordinary TM (the one that describes the workings of the UTM).

The great thing about the UTM is that it shows that all algorithms (Turing machines) can be reduced to a single algorithm. As stated above, Church, Gödel, and a number of other great thinkers did find alternative ways to represent algorithms, but it was only Turing who found a way of reducing all algorithms to a single one. This reduction in algorithms is a bit like what we have in information theory where all messages can be reduced to zeroes and ones.

2.3.4 *The Halting Problem*

This is a famous problem in computer science. Having discovered the simple but powerful model of computation (essentially the stored program computer), Turing then looked at its limitations by devising a problem that it could not solve.

The UTM can run any algorithm. What, asked Turing, if we have another UTM that, rather than running a given algorithm, looks to see whether that algorithm acting on its data will actually halt in a finite number of steps (rather than looping forever or crashing)? Turing called this hypothetical new Turing machine H (for halting machine). Like a UTM, H can receive a description of the algorithm in question (its program) and the algorithm's data. Then H works on this information and produces a result. When given a number, say 1 or 0 it decides whether or not the given algorithm would then halt. Turing asked if such a machine was possible. The answer he found was **no** (look below). The very concept of H involves a contradiction! He demonstrated this by taking a variant of H itself as both the algorithm description (program) and data that H should work on. This proof by contradiction only applies to Turing machines and machines that are computationally equivalent. It still remains unproven that the halting problem cannot be solved in *all* computational models.

The next section contains a detailed explanation of the halting problem by means of an example. This can be skipped if you've had enough of Turing machines for now.

2.3.4.1 *The Halting Problem — Proof by Contradiction*

Here we'll look at the halting problem in Javascript [26]. The proof is by contradiction: say we could have a program that can determine whether or not another program will halt.

```
function Halt(program) {
  if ( ...Code to check if program can halt... ) {
    return true;
  } else {
    return false;
  }
}
```

Given two programs, one that halts and one that does not:

```
function Halter(input) {
  alert('finished');
}


function Looper(input) {
  while (1==1) {;}
}
```

In our example Halt() would return the following:

```
Halt("function Halter(1){alert('finished');}")
    \\ returns  true
Halt("function Looper(1){while (1==1) {;}}")
    \\ returns  false
```

So it would be possible given these special cases, but is it possible for all algorithms to be covered in the ...Code to check if program can halt... section? No — given a new program:

```
function Contradiction(program) {
  if (Halt(program) == true) {
    while (1 == 1) {;}
  } else {
    alert('finished');
  }
}
```

If Contradiction() is given an arbitrary program as an input then:

- If Halt() returns true then Contradiction() goes into an infinite loop.
- If Halt() returns false then Contradiction() halts.

If Halt(\url{Contradiction()}) returns true then:

- Contradiction() loops infinitely if Halt(\url{Contradiction()}) halts.
- Contradiction() halts if Halt(\url{Contradiction()}) goes into an infinite loop.

`Contradiction()` here does not loop or halt, and we can't decide algorithmically what the behaviour of `Contradiction()` will be.

2.4 Circuits

Although modern computers are no more powerful than TM's they are a lot more efficient (for more on efficiency see section 2.5). However, what a modern or conventional computer gives in efficiency it loses in transparency (compared with a TM). It is for this reason that a TM is still of value in theoretical discussions, e.g. in comparing the hardness/difficulty of various classes of problems.

We won't go fully into the architecture of a conventional computer, but some of the concepts needed for quantum computing are related, e.g. circuits, registers, and gates. For this reason we'll examine conventional (classical) circuits.

Classical circuits are read left to right and made up of the following:

(1) *Gates* — which perform logical operations on inputs. Given input(s) with values of 0 or 1 they produce an output of 0 or 1 (see below). These operations can be represented by *truth tables* which specify all of the different combinations of the outputs with respect to the inputs.
(2) *Wires* — which carry signals between gates and registers.
(3) *Registers* — which are made up of cells containing 0 or 1, i.e. bits.

2.4.1 *Common Gates*

First off we'll look at the commonly used gates. These are listed below with their respective truth tables.

The `NOT` gate inverts the input bit. The gate part of the diagram is the triangle and circle, The wires are the lines on either side of the gate and the gate reads from left to right (as with all gates and circuits).

NOT	
a	x
0	1
1	0

OR returns a 1 if either of the inputs is 1.

OR		
a	b	x
0	0	0
1	0	1
0	1	1
1	1	1

AND only returns a 1 if both of the inputs are 1.

AND		
a	b	x
0	0	0
1	0	0
0	1	0
1	1	1

NOR is an OR and a NOT gate combined.

NOR		
a	b	x
0	0	1
1	0	0
0	1	0
1	1	0

NAND is an AND and a NOT gate combined.

NAND		
a	b	x
0	0	1
1	0	1
0	1	1
1	1	0

XOR returns a 1 if only one of its inputs is 1.

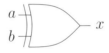

XOR		
a	b	x
0	0	0
1	0	1
0	1	1
1	1	0

2.4.2 Combinations of Gates

Shown below, a circuit also has a truth table.

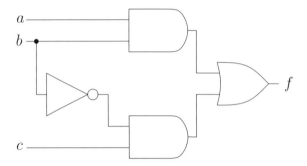

The circuit shown above has the following truth table:

a	b	c	f
0	0	0	0
0	0	1	1
0	1	0	0
0	1	1	0
1	0	0	0
1	0	1	1
1	1	0	1
1	1	1	1

This circuit has an expression associated with it, which is as follows:

$$f = \text{OR}(\text{AND}(a, b), \text{AND}(\text{NOT}(b), c)).$$

2.4.3 Relevant Properties

These circuits are capable of *FANOUT*, *FANIN*, and *CROSSOVER* (unlike quantum circuits) where FANOUT means that many wires (i.e. many inputs) can be tied to one output (figure 2.7), FANIN means that many outputs can be tied together with an OR, and CROSSOVER means that the value of two bits are interchanged.

2.4.4 Universality

Combinations of NAND gates can be used to emulate any other gate (see figure 2.8). For this reason the NAND gate is considered a universal gate. This means that any circuit, no matter how complicated can be expressed

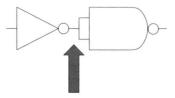

Fig. 2.7 FANOUT.

as a combination of NAND gates. The quantum analogue of this is called the CNOT gate.

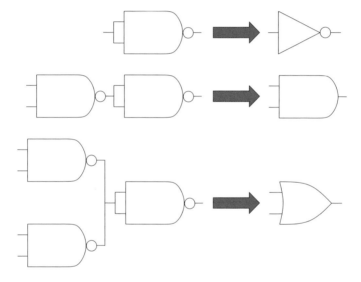

Fig. 2.8 The universal NAND gate.

2.5 Computational Resources and Efficiency

Computational time complexity is a measure of how fast and with how many resources a computational problem can be solved. In terms of algorithms, we can compare algorithms that perform the same task and measure whether one is more efficient than the other. Conversely, if the same algorithm is implemented on different architectures then the time complexities should not differ by more than a constant. This is called the *principle of invariance*.

A simple example of differing time complexities is with sorting algorithms, i.e. algorithms used to sort a list of numbers. The following example uses the bubble sort and quicksort algorithms. Some code for bubble sort is given below, but the code for quicksort is not given explicitly. We'll use the code for the bubble sort algorithm in an example on page 20.

Bubble sort:

```
for(i = 1;i < = n - 1;i ++)
  for(j = n;j > = i + 1;j --)
  {
    if (list[j]<list[j - 1])
    {
      Swap(list[j],list[j - 1]);
    }
  }
}
```

Example   Quicksort vs. bubble sort.

Bubble sort:
Each item in a list is compared with the item next to it, they are swapped if required. This process is repeated until all of the numbers are checked without any swaps.

Quicksort:
The quicksort has the following four steps:
(1) Finish if there are no more elements to be sorted (i.e. one or less elements).
(2) Select a pivot point.
(3) The list is split into two lists. Numbers smaller than the pivot value are in one list and numbers larger than the pivot value in the other list.
(4) Repeat steps 1 to 3 for each list generated in step 3.

On average the bubble sort is much slower than the quicksort, regardless of the architecture it is running on.

2.5.1 *Quantifying Computational Resources*

Let's say we've gone through an algorithm systematically and worked out, line by line, how fast it is going to run given a particular variable n which describes the size of the input. Suppose we can quantify the computational work involved as function of n. Consider the following expression:

$$3n + 2\log n + 12.$$

The important part of this function is $3n$ as it grows more quickly than the other terms. n grows faster than $\log n$ and the constant. We say that the algorithm that generated this result has,

$$O(n)$$

time complexity (i.e. we ignore the 3). The important parts of the function are shown here:

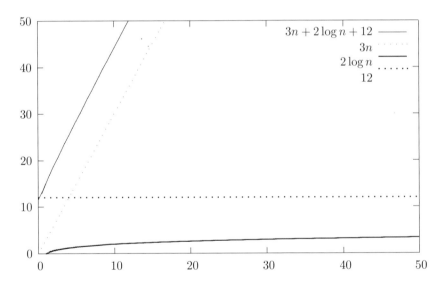

Here we have split the function $3n + 2\log n + 12$ into its parts: $3n$, $2\log n$, and 12.

More formally *Big O* notation allows us to set an upper bound on the behaviour of the algorithm. So, at worst this algorithm will take approximately n cycles to complete (plus a vanishingly small unimportant figure).

Note that this is the *worst case*, i.e. it gives us no notion of the average complexity of an algorithm. The class of $O(n)$ contains all functions that are *quicker* than $O(n)$.

A more formal definition of Big O notation is as follows:

$$f(n) \text{ is } O(g(n)) \text{ if there exist constants } n_0 \text{ and } c$$
$$\text{that for all } n > n_0, \ f(n) \leq cg(n). \tag{2.2}$$

Here is a more in-depth example of the average complexity of an algorithm.

Example Time Complexity: Quicksort vs. bubble sort [31].

Here if n is the number of elements in a list, the number of swap operations averages to:

$$(n-1) + (n-2) + ... + 1 = \frac{n(n-1)}{2}$$

The most important factor here is n^2. The average and worst case time complexities are $O(n^2)$, so we say it is generally $O(n^2)$. If we do the same to the quicksort algorithm, the average time complexity is just $O(n \log n)$.

So now we have a precise mathematical notion for the speed of an algorithm. Note: there are other commonly used related notations like big Ω and big Θ but big O is all we'll need for now.

2.5.2 *Standard Complexity Classes*

Computational complexity is the study of how hard a problem is to compute. Or put another way, what the least amount of resources required by the best known algorithm for solving the problem is. There are many types of resources (e.g. time and space), but we are interested in only time complexity for now. The main distinction between hard and easy problems is the rate at which they grow. If the problem can be solved in *polynomial* time, that is, it is bounded by a polynomial in n, then it is easy. Hard problems grow faster than any polynomial in n, for example:

$$n^2$$

is polynomial, and easy, whereas,

$$2^n$$

is *exponential* and hard.

What we mean by hard is that as we make n large the time taken to solve the problem goes up as 2^n, i.e. exponentially. So we say that $O(2^n)$ is hard or *intractable*.

The complexity classes of most interest to us are **P** (Polynomial) and **NP** (Nondeterministic Polynomial). **P** means that the problem can be solved in polynomial time, **NP** means that it probably can't, and **NP** complete means it almost definitely can't. More formally:

> The class **P** consists of all those decision problems that can be solved on a deterministic sequential machine in an amount of time that is polynomial in the size of the input; the class **NP** consists of all those decision problems whose positive solutions can be verified in polynomial time given the right information, or equivalently, whose solution can be found in polynomial time on a nondeterministic machine [quote taken from *Wikipedia*].

Where:

- A *decision problem* is a function that takes an arbitrary value or values as input and returns a yes or a no. Most problems can be represented as decision problems. Solutions to decision problems that can be checked in polynomial time are called witnesses. For example, checking if 7747 has a factor less than 70 is a decision problem. 61 is a factor ($61 \times 127 = 7747$) which is easily checked. This is an example of a witness with a positive solution we can check in polynomial time. Now, if we ask if 7747 has a factor less than 60 there is no witnesses for the no instance so we have to check every number between 60 and 2 for a positive solution [30] before answering no.
- *Nondeterministic* Turing Machines (NTMs) differ from normal *deterministic* Turing machines in that at each step of the computation the Turing machine can "spawn" copies, or new Turing machines that work in parallel with the original. It's a common mistake to call a quantum computer an NTM, as we shall see later we can only use quantum parallelism indirectly.

- It is not proven that $\mathbf{P} \neq \mathbf{NP}$; it is just very unlikely as this would mean that all problems in $\mathbf{NP}$ can be solved in polynomial time.

You can also see a large list of complexity classes online at `http://qwiki.stanford.edu/index.php/Complexity_Zoo`.

2.5.3 *The Strong Church–Turing Thesis*

Originally, the strong Church–Turing thesis went something like this:

> Any algorithmic process can be simulated *with no loss of efficiency* using a Turing machine [2].

This is saying a TM is as powerful as any other model of computation in terms of the class of problems it can solve; any efficiency gain due to using a particular model is at most polynomial.

This was challenged in 1977 by Robert Solovay and Volker Strassen, who introduced truly randomised algorithms which do give a computational advantage based on the machine's architecture [31]. So, this led to a revision of the strong Church–Turing thesis, which now relates to a *Probabilistic Turing Machine* (PTM).

A probabilistic Turing machine can be described as:

> A deterministic Turing machine having an additional write instruction where the value of the write is uniformly distributed in the Turing machine's alphabet (generally, an equal likelihood of writing a 1 or a 0 on to the tape) [quote taken from *The Free Dictionary*].

An example of an algorithm that can benefit from a PTM is quicksort. Although on average quicksort runs in $O(n \log n)$ it still has a worst case running time of $O(n^2)$ if the list is already sorted. Randomising the list beforehand ensures the algorithm runs in $O(n \log n)$ more often.

Can we efficiently simulate any nonprobabilistic algorithm on a probabilistic Turing machine without exponential slowdown? The answer is yes according to the new strong Church–Turing thesis:

> Any model of computation can be simulated on a probabilistic Turing machine with at most a polynomial increase in the number of elementary operations [5].

A new challenge came from another quarter when in the early eighties when Richard Feynman, 1918–1988 (figure 2.9) suggested that it would

Fig. 2.9   Richard Feynman.

be possible to simulate quantum systems using quantum mechanics. This alluded to a kind of proto-quantum computer. He then went on to ask if it was possible to simulate quantum systems on conventional (i.e. classical) Turing machines. It is hard to simulate quantum systems effectively, in fact it gets exponentially harder the more *components* you have [30]. Intuitively, the TM simulation can't keep up with the evolution of the physical system itself: it falls further and further behind, exponentially so. Then, reasoned Feynman, if the simulator was built of quantum components perhaps it wouldn't fall behind. So such a quantum computer would seem to be more efficient than a TM. The strong Church–Turing thesis would seem to have been violated (as the two models are not polynomially equivalent).

The idea really took shape in 1985 when, based on Feynman's ideas, David Deutsch proposed another revision to the strong Church Turing thesis. He proposed a new architecture based on quantum mechanics, on the assumption that all physics is derived from quantum mechanics. This is called the *Deutsch–Church–Turing principle* [30]. He then demonstrated a simple quantum algorithm which seemed to prove the new revision. More algorithms were developed that seemed to work better on a quantum Turing machine (see below) than a classical one, notably Shor's factorisation and Grover's search algorithms which are described in chapter 7.

2.5.4   *Quantum Turing Machines*

A *Quantum Turing Machine* (QTM) is a normal Turing machine with quantum parallelism. The head and tape of a QTM exist in quantum states, and each cell of the tape holds a quantum bit which can contain what's called a superposition of the values 0 and 1. Don't worry too much about that now as it'll be explained in detail later; what's important is that a

QTM can perform calculations on a number of values simultaneously by using quantum effects. Classical parallelism requires a separate processor for each value operated on in parallel but in quantum parallelism a single processor operates on all the values simultaneously.

Question   *Architecture itself can change the time complexity of algorithms so could there be other revisions?*

2.6   Energy and Computation

2.6.1   *Reversibility*

When an isolated quantum system evolves it always does so *reversibly*. At any time you could determine an earlier state of the system by using its current state and "playing the system backwards". This implies that if a quantum computer has components that perform logical operations then these components will have to implement logical operations reversibly.

2.6.2   *Irreversibility*

Most classical circuits are not reversible. This means that they lose information in the process of generating outputs from inputs, i.e. they are not *invertible*. An example of this is the NAND gate (figure 2.10). It is not possible in general to invert the output. For example, knowing the output is 1 does not allow one to determine the input: it could be $00, 10,$ or 01.

Fig. 2.10   An irreversible NAND gate.

2.6.3   *Landauer's Principle*

In 1961, IBM physicist Rolf Landauer, 1927–1999 showed that when information is lost in an irreversible circuit that information is dissipated as heat [30]. This result was obtained for circuits based on classical physics.

Theoretically, if we were to build a classical computer with reversible components then work could be done with no heat loss, and no use of energy! Practically though we still need to waste some energy for correcting any physical errors that occur during the computation. A good example of the link between reversibility and information is *Maxwell's demon*, which is described next.

2.6.4 *Maxwell's Demon*

Maxwell's demon is a thought experiment comprised of (see figure 2.11) a box filled with gas separated into two halves by a wall. The wall has a little door that can be opened and closed by a demon. The second law of thermodynamics (see chapter 4) says that the amount of entropy in a closed system never decreases. Entropy is the amount of disorder in a system or in this case the amount of energy. The demon can, in theory, open and close the door in a certain way to actually decrease the amount of entropy in the system.

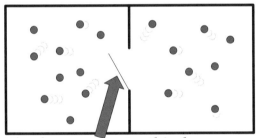

The demon operates this door

Fig. 2.11   Maxwell's Demon.

Here are a list of steps to understanding the problem:

(1) We have a box filled with particles that have different velocities (shown by the arrows).
(2) A demon opens and closes a door in the centre of the box that allows particles to travel through it.
(3) The demon only opens the door when fast particles come from the right and slow ones from the left.
(4) The fast particles end up on the left hand side, the slow particles on the right. The demon makes a temperature difference without doing any work (which violates the second law of thermodynamics).

(5) Rolf Landauer and R.W. Keyes resolved the paradox when they examined the thermodynamic costs of information processing. The demon's mind gets hotter as his memory stores the results. The operations are reversible until his memory is cleared.

(6) Almost anything can be done in a reversible manner (with no entropy cost).

2.6.5   *Reversible Computation*

In 1973 Charles Bennett expanded on Landauer's work and asked whether it was possible, in general, to do computational tasks without dissipating heat. The loss of heat is not important to quantum circuits, but because quantum mechanics is reversible we must build quantum computers with *reversible gates*.

We can simulate any classical gate with reversible gates. For example, a reversible NAND gate can be made from a reversible gate called a *Toffoli* gate.

Reversible gates use *control lines* which in reversible circuits can be fed from *ancilla bits* (which are work bits). Bits in reversible circuits may then go on to become *garbage bits* that are only there to ensure reversibility. Control lines ensure we have enough bits to recover the inputs from the outputs. The reason they are called control lines is that they control (as in an if statement) whether or not a logic operation is applied to the non-control bit(s). E.g. in CNOT below, the NOT operation is applied to bit b if the control bit is on (=1).

2.6.6   *Reversible Gates*

Listed below are some of the common reversible gates and their truth tables. Note: the reversible gate diagrams, and quantum circuit diagrams were built with a LaTeX macro package called *Q-Circuit* which is available at http://info.phys.unm.edu/Qcircuit/.

2.6.6.1   *Controlled NOT*

Like a NOT gate (on b) but with a control line, a. b' can also be expressed as a XOR b.

CNOT			
a	b	a'	b'
0	0	0	0
0	1	0	1
1	0	1	1
1	1	1	0

Properties of the CNOT gate, CNOT(a, b):

$$\text{CNOT}(x, 0) : b' = a' = a = \text{FANOUT}. \tag{2.3}$$

2.6.6.2 *Toffoli Gate*

If the two control lines are set it flips the third bit (i.e. applies NOT). The Toffoli gate is also called a controlled-controlled NOT.

Toffoli					
a	b	c	a'	b'	c'
0	0	0	0	0	0
0	0	1	0	0	1
0	1	0	0	1	0
0	1	1	0	1	1
1	0	0	1	0	0
1	0	1	1	0	1
1	1	0	1	1	1
1	1	1	1	1	0

Properties of the Toffoli Gate, $T_F(a, b, c)$:

$$T_F(a, b, c) = (a, b, c \text{ XOR}(a \text{ AND } b)). \tag{2.4}$$

$$T_F(1, 1, x) : c' = \text{NOT } x. \tag{2.5}$$

$$T_F(x, y, 1) : c' = x \text{ NAND } y. \tag{2.6}$$

$$T_F(x, y, 0) : c' = x \text{ AND } y. \tag{2.7}$$

$$T_F(x, 1, 0) : c' = a = a' = \text{FANOUT}. \tag{2.8}$$

A combination of Toffoli gates can simulate a *Fredkin* gate.

2.6.6.3   *Fredkin Gate*

If the control line is set it flips the second and third bits.

Fredkin					
a	b	c	a'	b'	c'
0	0	0	0	0	0
0	0	1	0	0	1
0	1	0	0	1	0
0	1	1	0	1	1
1	0	0	1	0	0
1	0	1	1	1	0
1	1	0	1	0	1
1	1	1	1	1	1

Properties of the Fredkin Gate, $F_R(a, b, c)$:

$$F_R(x, 0, y) : b' = x \text{ AND } y. \tag{2.9}$$

$$F_R(1, x, y) : b' = c \text{ and } c' = b, \text{ which is CROSSOVER}. \tag{2.10}$$

$$F_R(x, 1, 0) : c' = a' = c = \text{FANOUT, with } b' = \text{NOT } x. \tag{2.11}$$

A combination of Fredkin gates can simulate a Toffoli gate.

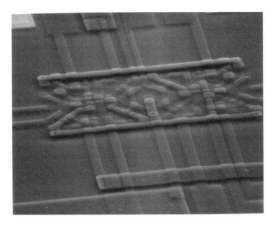

Fig. 2.12 A conventional reversible circuit.

2.6.7 *Reversible Circuits*

Reversible circuits have been implemented in a classical sense. An example of a reversible circuit built with conventional technology is shown in figure 2.12. Quantum computers use reversible circuits to implement quantum algorithms. Chapters 6 and 7 contain many examples of these algorithms and their associated circuits.

Chapter 3

Mathematics for Quantum Computing

3.1 Introduction

In conventional computers we have logical operators (gates) such as `NOT` that acts on bits. The quantum analogue of this is a matrix operator operating on a qubit state vector. The mathematics we need to handle this includes:

- Vectors to represent the quantum state.
- Matrices to represent gates acting on the values.
- Complex numbers, because the components of the quantum state vector are in general complex.
- Trig functions for the polar representation of complex numbers and the Fourier series.
- Projectors to handle quantum measurements.
- Probability theory for computing the probability of measurement outcomes.

As well as there being material here that you may not be familiar with (complex vector spaces for example), chances are that you'll know at least some of the mathematics. The sections you know might be useful for revision, or as a reference. This is especially true for the sections on polynomials, trigonometry, and logs which are very succinct.

So what's not in here? There's obviously some elementary mathematics that is not covered. This includes topics like fractions, percentages, basic algebra, powers, radicals, summations, limits, factorisation, and simple geometry. If you're not comfortable with these topics then you may need to study them before continuing on with this chapter.

3.2 Polynomials

A *polynomial* is an expression in the form:

$$c_0 + c_1 x + c_2 x^2 + ... + c_n x^n \qquad (3.1)$$

where $c_0, c_1, c_2, ..., c_n$ are *constant coefficients* with $c_n \neq 0$.

We say that the above is a polynomial in x of degree n.

Example   Different types of polynomials.

$3v^2 + 4v + 7$ is a polynomial in v of degree 2, i.e. a quadratic.
$4t^3 - 5$ is a polynomial in t of degree 3, i.e. a cubic.
$6x^2 + 2x^{-1}$ is not a polynomial as it contains a negative power for x.

3.3 Logical Symbols

A number of logical symbols are used in this text to compress formulae; they are explained below:

$\forall$ means *for all.*

Example $\forall\, n > 5, f(n) = 4$ means that for all values of n greater than 5, $f(n)$ will return 4.

$\exists$ means *there exists.*

Example $\exists\, n$ such that $f(n) = 4$ means there is a value of n that will make $f(n)$ return 4. Say if $f(n) = (n - 1)^2 + 4$, then the n value in question is $n = 1$.

---

iff means *if and only if.*

Example $f(n) = 4$ iff $n = 8$ means $f(n)$ will return 4 if $n = 8$ but for no other values of n.

3.4   Trigonometry Review

3.4.1   *Right Angled Triangles*

Given the triangle,

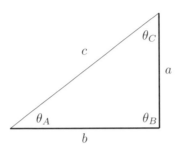

we can say the following:

$$a^2 + b^2 = c^2 \; , \qquad\qquad\qquad Pythagorean\ theorem \quad (3.2)$$

and for the **opp**osite side, **adj**acent side, and **hyp**otenuse:

$$\sin = \frac{\mathbf{opp}}{\mathbf{hyp}} \; , \qquad \cos = \frac{\mathbf{adj}}{\mathbf{hyp}} \; , \qquad \tan = \frac{\mathbf{opp}}{\mathbf{adj}} \; , \qquad\qquad (3.3)$$

$$\sin \theta_A = \frac{a}{c} \; , \qquad \sin \theta_B = \frac{b}{c} \; , \qquad\qquad (3.4)$$

$$\tan \theta_A = \frac{a}{b} \; , \qquad \tan \theta_B = \frac{b}{a} \; , \qquad\qquad (3.5)$$

$$\cos \theta_A = \sin \theta_B = \frac{b}{c} \; , \qquad \cos \theta_B = \sin \theta_A = \frac{a}{c} \; . \qquad (3.6)$$

3.4.2   *Converting Between Degrees and Radians*

Angles in trigonometry can be represented in radians and degrees. For converting degrees to radians:

$$\text{rads} = \frac{n^\circ \times \pi}{180} \; . \qquad\qquad (3.7)$$

For converting radians to degrees we have:

$$n° = \frac{180 \times \text{rads}}{\pi} . \tag{3.8}$$

Some common angle conversions are:

$360° = 0° = 2\pi$ rads.

$1° = \frac{\pi}{180}$ rads.

$45° = \frac{\pi}{4}$ rads.

$90° = \frac{\pi}{2}$ rads.

$180° = \pi$ rads.

$270° = \frac{3\pi}{2}$ rads.

1 rad $\approx 57°$.

3.4.3 *Inverses*

Here are some inverses for obtaining θ from $\sin\theta, \cos\theta,$ and $\tan\theta$:

$$\sin^{-1} = \arcsin = \theta \text{ from } \sin\theta. \tag{3.9}$$

$$\cos^{-1} = \arccos = \theta \text{ from } \cos\theta. \tag{3.10}$$

$$\tan^{-1} = \arctan = \theta \text{ from } \tan\theta. \tag{3.11}$$

3.4.4 *Angles in Other Quadrants*

The angles for right angled triangles are in quadrant 1 (i.e. from $0°$ to $90°$). If we want to measure larger angles like $247°$ we must determine which quadrant the angle is in (here we don't consider angles larger than $360°$). The following diagram has the rules for doing so:

Change θ to $180° - \theta$ Make cos and tan negative	No change
$90° \leq \theta \leq 180°$	$0° \leq \theta \leq 90°$
$180° \leq \theta \leq 270°$	$270° \leq \theta \leq 360°(0°)$
Change θ to $\theta - 180°$ Make sin and cos negative	Change θ to $360° - \theta$ Make sin and tan negative

Example Using the diagram above we can say that $\sin(315°) = -\sin(45°)$ and $\cos(315°) = \cos(45°)$.

3.4.5 *Visualisations and Identities*

The functions $y = \sin(x)$ and $y = \cos(x)$ are shown graphically below, where x is in radians.

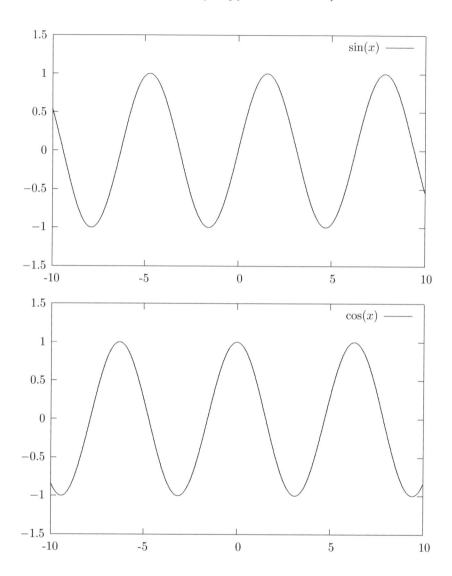

Finally, here are some important identities (where $\sin^2 \theta = (\sin \theta)^2$ and $\cos^2 \theta = (\cos \theta)^2$):

$$\sin^2 \theta + \cos^2 \theta = 1. \tag{3.12}$$

$$\sin(-\theta) = -\sin \theta. \tag{3.13}$$

$$\cos(-\theta) = \cos \theta. \tag{3.14}$$
$$\tan(-\theta) = -\tan \theta. \tag{3.15}$$

3.5 Logs

The *logarithm* of a number (say x) to base b is the power of b that gives back the number, i.e. $b^{\log_b x} = x$. For example, the log of $x = 100$ to base $b = 10$ is the power (2) of 10 that gives back 100, i.e. $10^2 = 100$. So $\log_{10} 100 = 2$.

Put another way, the answer to a logarithm is the power y put to a base b given an answer x, with:

$$y = \log_b x \tag{3.16}$$

and,

$$x = b^y \tag{3.17}$$

where $x >= 0$, $b >= 0$, and $b \neq 1$.

Example $\log_2 16 = 4$ is equivalent to $2^4 = 16$.

3.6 Complex Numbers

A *complex number*, z is a number in the form:

$$z = a + ib \tag{3.18}$$

where $a, b \in \mathbb{R}$ (the real numbers) and i stands for $\sqrt{-1}$. The complex number z is said to be in $\mathbb{C}$ (the complex numbers). z is called complex because it is made of two parts, a and b. Sometimes we write $z = (a, b)$ to express this. i is called an *imaginary* number and was created to solve problems like $x^2 + 1 = 0$, which had no traditional solution.

Imaginary axis

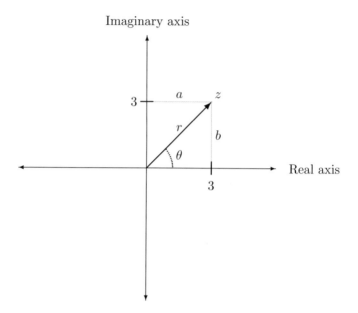

Fig. 3.1   Representing $z = a + ib$ in the complex plane with coordinates $a = 3$ and $b = 3$. Note the imaginary axis which accounts for i.

Except for the rules regarding i, the operations of addition, subtraction, and multiplication of complex numbers follow the normal rules of arithmetic. Division requires using a complex conjugate, which is introduced in the next section. These operations are defined via the examples in the box below.

The system of complex numbers is closed in that, except for division by 0, sums, products, and ratios of complex numbers give back a complex number: i.e. we stay within the system. Here are examples of i itself:

$$i^{-3} = i, \quad i^{-2} = -1, \quad i^{-1} = -i, \quad i = \sqrt{-1},$$

$$i^2 = -1, \quad i^3 = -i, \quad i^4 = 1, \quad i^5 = i, \quad i^6 = -1.$$

So the pattern $(-i, 1, i, -1)$ repeats indefinitely (aside from $i = \sqrt{-1}$).

Example   Basic complex numbers.

Addition:

$$(5 + 2i) + (-4 + 7i) = 1 + 9i$$

Multiplication:

$$(5 + 2i)(-4 + 3i) = 5(-4) + 5(3)i + 2(-4)i + (2)(3)i^2$$
$$= -20 + 15i - 8i - 6$$
$$= -26 + 7i.$$

Finding Roots:

$$(-5i)^2 = (-5i)(-5i)$$
$$= 25i^2$$
$$= 25(-1)$$
$$= -25.$$

-25 has roots $5i$ and $-5i$.

3.6.1   *Polar Coordinates and Complex Conjugates*

Complex numbers can be represented in *polar form*, (r, θ) where r is the magnitude and θ is called the phase:

$$(r, \theta) = (|z|, \theta) = |z|(\cos \theta + i \sin \theta) \tag{3.19}$$

where $\theta, r \in \mathbb{R}$ and $|z|$ is the *norm* (also called the *modulus*) of z:

$$|z| = \sqrt{a^2 + b^2} \tag{3.20}$$

or,

$$|z| = \sqrt{z^* z} \tag{3.21}$$

where z^* is the *complex conjugate* of z:

$$z^* = a - ib. \tag{3.22}$$

Two complex numbers may be equivalent when they differ in phase. Complex numbers are only unique if the phase is restricted to $0 \leq \theta \leq 2\pi$. It is not practical to restrict the phase during calculations (we may for example want to add angles) but afterwards we can reduce the phase modulo 2π. For two polar representations of complex numbers, if the magnitude is the same and the the phase is the same modulo 2π then they are equivalent.

Example   Reduction by modulo 2π.

$(2, \pi)$ and $(2, 5\pi)$ are equivalent because the phase differs by $(2)2\pi$.

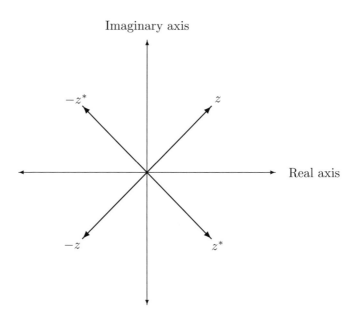

Fig. 3.2 z, z^*, $-z^*$, and $-z$.

3.6.1.1 *Polar Coordinates*

A complex number can be represented as a point on the *complex plane*. As well as being expressed by cartesian coordinates the point can be expressed using *polar coordinates* (figure 3.1). The angle θ is the angle between a line drawn from the origin to point (a, b) and the x axis. This line has length r. The horizontal axis is called the real axis and the vertical axis is called the *imaginary axis*. It's also helpful to look at the relationships between z, z^*, $-z^*$ and $-z$ graphically. These are shown in figure 3.2.

So for converting from polar to cartesian coordinates:

$$(r, \theta) = a + bi \qquad (3.23)$$

where $a = r\cos\theta$ and $b = r\sin\theta$. Conversely, converting cartesian to polar form is a little more complicated:

$$a + bi = (r, \theta) \qquad (3.24)$$

where $r = |z| = \sqrt{a^2 + b^2}$ and θ is the solution to $\tan\theta = \frac{b}{a}$ which lies in the following quadrant:

(1) If $a > 0$ and $b > 0$
(2) If $a < 0$ and $b > 0$
(3) If $a < 0$ and $b < 0$
(4) If $a > 0$ and $b < 0$.

Example   Convert $(3, 40°)$ to $a + bi$.

$$a = r \cos \theta = 3 \cos 40°$$
$$= 3(0.77)$$
$$= 2.3$$

$$b = r \sin \theta = 3 \cos 40°$$
$$= 3(0.64)$$
$$= 1.9$$

$$z = 2.3 + 1.9i \; .$$

Example   Convert $-1 + 2i$ to (r, θ). This gives us $a = -1$ and $b = 2$.

$$r = \sqrt{(-1)^2 + 2^2}$$
$$= \sqrt{5}$$
$$= 2.2$$

$$\tan \theta = \frac{b}{a}$$
$$= \frac{2}{-1}$$
$$= -2$$

$$\arctan(-2) = -63.4° \; .$$

Since $a < 0$ and $b > 0$ we use quadrant 2 so we negate -63.4 and subtract that from 180 which gives us $\theta = 116.6°$. The solution is:

$$-1 + 2i = (2.2, 116.6°) \; .$$

3.6.2 *Rationalising and Dividing*

$\frac{1}{a+bi}$ is *rationalised* by multiplying the numerator and denominator by $a-bi$.

Example Rationalisation.

$$\frac{1}{5+2i} = \frac{1}{5+2i}\frac{(5-2i)}{(5-2i)}$$

$$= \frac{5}{29} - \frac{2}{29}i \ .$$

Division of complex numbers is done by rationalising in terms of the denominator.

Example Division of complex numbers.

$$\frac{3+2i}{2i} = \frac{3+2i}{2i}\frac{(-2i)}{(-2i)}$$

$$= \frac{-6i-4i^2}{-4i^2}$$

$$= \frac{-6i+4}{4}$$

$$= 1 - \frac{3}{2}i \ .$$

3.6.3 *Exponential Form*

Complex numbers can also be represented in *exponential form*:

$$z = re^{i\theta}. \tag{3.25}$$

The derivation of which is:

$$z = |z|(\cos\theta + i\sin\theta)$$
$$= r(\cos\theta + i\sin\theta)$$
$$= re^{i\theta}.$$

This is because:

$$e^{i\theta} = \cos\theta + i\sin\theta, \tag{3.26}$$

$$e^{-i\theta} = \cos\theta - i\sin\theta. \tag{3.27}$$

which can be rewritten as:

$$\cos\theta = \tfrac{e^{i\theta} + e^{-i\theta}}{2}, \tag{3.28}$$

$$\sin\theta = \tfrac{e^{i\theta} - e^{-i\theta}}{2i}. \tag{3.29}$$

To prove (3.26) we use a power series exponent (which is an infinite polynomial):

$$e^{x} = 1 + x + \frac{x^2}{2!} + \frac{x^3}{3!} + \dots, \tag{3.30}$$

$$e^{i\theta} = 1 + i\theta - \frac{\theta^2}{2!} - \frac{i\theta^3}{3!} + \frac{\theta^4}{4!} - \dots \tag{3.31}$$

$$= 1 - \frac{\theta^2}{2!} + \frac{\theta^4}{4!} + i(\theta - \frac{i\theta^3}{3!} + \dots) \tag{3.32}$$

$$= \cos\theta + i\sin\theta. \tag{3.33}$$

Example Convert $3 + 3i$ to exponential form. This requires two main steps, which are:

(1) Find the modulus.

$$r = |z| = \sqrt{3^2 + 3^2}$$

$$= \sqrt{18}.$$

(2) To find θ, we can use the a and b components of z as opposite and adjacent sides of a right angled triangle in quadrant one (see figure 3.1) which means we need to apply arctan. So given $\tan^{-1} \frac{3}{3} = \frac{\pi}{4}$ then z in exponential form looks like:

$$\sqrt{18}e^{\pi i/4} \ .$$

Example Convert $e^{\pi i 3/4}$ to the form: $a + bi$ (also called rectangular form).

$$e^{\pi i 3/4} = e^{i(3\pi/4)}$$

$$= \cos \frac{3\pi}{4} + i \sin \frac{3\pi}{4}$$

$$= \cos 135° + i \sin 135°$$

$$= \frac{-1}{\sqrt{2}} + \frac{i}{\sqrt{2}}$$

$$= \frac{-1 + i}{\sqrt{2}} \ .$$

Properties:

$$z^* = re^{-i\theta}. \tag{3.34}$$

$$e^{-i2\pi} = 1. \tag{3.35}$$

3.7 Matrices

Matrices will be needed in quantum computing to represent gates, operators, and vectors. So even if you know this material it'll be useful to revise as they are used so often.

A matrix is an array of numbers, the numbers in the matrix are called *entries*, for example:

$$\begin{bmatrix} 17 & 24 & 1 & 8 \\ 23 & 5 & 7 & 14 \\ 4 & 6 & 13 & 20 \end{bmatrix}.$$

3.7.1   Matrix Operations

Just as we could define arithmetic operators — addition and multiplication for complex numbers, we can do the same for matrices.

Given the following 3 matrices:

$$M_A = \begin{bmatrix} 2 & 1 \\ 3 & 4 \end{bmatrix},$$

$$M_B = \begin{bmatrix} 2 & 1 \\ 3 & 5 \end{bmatrix},$$

$$M_C = \begin{bmatrix} 2 & 1 & 0 \\ 3 & 4 & 0 \end{bmatrix}.$$

3.7.1.1   Addition

Addition can only be done when the matrices are of the same dimensions (the same number of columns and rows), e.g:

$$M_A + M_B = \begin{bmatrix} 4 & 2 \\ 6 & 9 \end{bmatrix}.$$

3.7.1.2   Scalar Multiplication

The product of multiplying a *scalar* (i.e. a number) by a matrix is a new matrix that is found by multiplying each entry in the given matrix. Given a scalar $\alpha = 2$:

$$\alpha M_A = \begin{bmatrix} 4 & 2 \\ 6 & 8 \end{bmatrix}.$$

3.7.1.3 *Matrix Multiplication*

The product of multiplying matrices M and N with dimensions $M = m \times r$ and $N = r \times n$ is a matrix O with dimension $O = m \times n$. The resulting matrix is found by $O_{ij} = \sum_{k=1}^{r} M_i r N_r j$ where i and j denote row and column respectively. The matrices M and N must also satisfy the condition that the number of columns in M is the same as the number of rows in N.

$$M_B M_C = \begin{bmatrix} (2 \times 2) + (1 \times 3) & (2 \times 1) + (1 \times 4) & (2 \times 0) + (1 \times 0) \\ (3 \times 2) + (5 \times 3) & (3 \times 1) + (5 \times 4) & (3 \times 0) + (5 \times 0) \end{bmatrix}$$

$$= \begin{bmatrix} 7 & 6 & 0 \\ 21 & 23 & 0 \end{bmatrix}.$$

3.7.1.4 *Basic Matrix Arithmetic*

Suppose M, N, and O are matrices and α and β are scalars:

$M + N = N + M.$	*Commutative law for addition*	(3.36)
$M + (N + O) = (M + N) + O.$	*Associative law for addition*	(3.37)
$M(NO) = (MN)O.$	*Associative law for multiplication*	(3.38)
$M(N + O) = MN + MO.$	*Distributive law*	(3.39)
$(N + O)M = NM + OM.$	*Distributive law*	(3.40)
$M(N - O) = MN - MO.$		(3.41)
$(N - O)M = NM - OM.$		(3.42)
$\alpha(N + O) = \alpha N + \alpha O.$		(3.43)
$\alpha(N - O) = \alpha N - \alpha O.$		(3.44)
$(\alpha + \beta)O = \alpha O + \beta O.$		(3.45)
$(\alpha - \beta)O = \alpha O - \beta O.$		(3.46)
$(\alpha\beta)O = \alpha(\beta O).$		(3.47)
$\alpha(NO) = (\alpha N)O = N(\alpha O).$		(3.48)

You may have noticed that there is no commutative law for multiplication. It is **not** always the case that $MN = NM$. This is important in quantum mechanics, which follows the same noncommutative multiplication law.

3.7.1.5 Zero Matrix

The special case of a matrix filled with zeroes.

$$\mathbf{0} = \begin{bmatrix} 0 & 0 \\ 0 & 0 \end{bmatrix}. \tag{3.49}$$

3.7.1.6 Identity Matrix

A matrix multiplied by the identity matrix (corresponding to unity in the ordinary numbers) will not change.

$$I = \begin{bmatrix} 1 & 0 \\ 0 & 1 \end{bmatrix}, \tag{3.50}$$

$$M_A I = \begin{bmatrix} 2 & 1 \\ 3 & 4 \end{bmatrix}.$$

3.7.1.7 Inverse Matrix

A number a has an inverse a^{-1} where $aa^{-1} = a^{-1}a = 1$. Equivalently a matrix A has an inverse:

$$A^{-1} \text{ where } AA^{-1} = A^{-1}A = I. \tag{3.51}$$

Even with a simple 2×2 matrix it is not a trivial matter to determine its inverses (if it has any at all). An example of an inverse is below, for a full explanation of how to calculate an inverse you'll need to consult an external reference.

$$M_A^{-1} = \begin{bmatrix} \frac{4}{5} & \frac{-1}{5} \\ \frac{-3}{5} & \frac{2}{5} \end{bmatrix}.$$

Note A^{-1} only exists iff A has full rank (see Determinants and Rank below).

3.7.1.8 *Transpose Matrix*

A^T is the *transpose* of matrix A if:

$$A^T_{ji} = A_{ij} .\tag{3.52}$$

Here's an example:

$$M^T_C = \begin{bmatrix} 2 & 3 \\ 1 & 4 \\ 0 & 0 \end{bmatrix}.$$

For a square matrix like M_A you get the transpose by reflecting about the diagonal (i.e. flipping the values).

3.7.1.9 *Determinants and Rank*

Rank is the number of rows (or columns) which are not linear combinations (see section 3.8.6) of other rows.

In the case of a square matrix A (i.e., $m = n$), A is invertible iff A has rank n (we say that A has full rank). A matrix has *full rank* when the rank is the number of rows or columns, whichever is smaller. A nonzero *determinant* (see below) determines that the matrix has full rank, so a nonzero determinant implies the matrix has an inverse and vice-versa. If the determinant is 0 the matrix is *singular* (i.e. doesn't have an inverse).

The determinant of a simple 2×2 matrix is defined as:

$$\det \left| \begin{bmatrix} a & b \\ c & d \end{bmatrix} \right| = ad - bc.\tag{3.53}$$

So now for an example of rank, given the matrix below,

$$M_D = \begin{bmatrix} 2 & 4 \\ 3 & 6 \end{bmatrix}.$$

We can say it has rank 1 because row 2 is a multiple (by $\frac{3}{2}$) of row 1. It has a determinant, $2 \times 6 - 3 \times 4$, of 0.

Determinants of larger matrices can be found by decomposing them into smaller 2×2 matrices, for example:

$$\det\left(\begin{bmatrix} a & b & c \\ d & e & f \\ g & h & i \end{bmatrix}\right) = a \cdot \det\left(\begin{bmatrix} e & f \\ h & i \end{bmatrix}\right) - b \cdot \det\left(\begin{bmatrix} d & f \\ g & i \end{bmatrix}\right) + c \cdot \det\left(\begin{bmatrix} d & e \\ g & h \end{bmatrix}\right). \quad (3.54)$$

Determinants, like inverses are not trivial to calculate. Again, for a full explanation you'll need to consult an external reference.

3.8 Vectors and Vector Spaces

3.8.1 *Introduction*

Vectors are line segments that have both magnitude and direction. Vectors for quantum computing are in the complex vector space $\mathbb{C}^n$ called n dimensional *Hilbert space*. But it's helpful to look at simpler vectors in real space (i.e. ordinary 2D space) first.

3.8.1.1 *Vectors in* $\mathbb{R}$

A vector in $\mathbb{R}$ (the real numbers) can be represented by a point on the cartesian plane (x, y) if the tail of the vector starts at the origin (see figure 3.3). The x and y coordinates that relate to the x and y axes are called the *components* of the vector.

The tail does not have to start at the origin and the vector can move anywhere in the cartesian plane as long as it keeps the same direction and length. When vectors do not start at the origin they are made up of two points, the *initial point* and the *terminal point*. For simplicity's sake our vectors in $\mathbb{R}$ all have an initial point at the origin and our coordinates just refer to the terminal point.

The collection of all the vectors corresponding to all the different points in the plane make up the space ($\mathbb{R}^2$). We can make a vector 3D by using another axis (the z axis) and extending into *3 space* ($\mathbb{R}^3$) (see figure 3.7). This can be further extended to more dimensions using *n space* ($\mathbb{R}^n$).

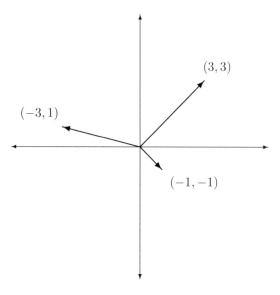

Fig. 3.3   Vectors in $\mathbb{R}^2$ (i.e. ordinary 2D space like a table top).

Example   A point in 5 dimensional space is represented by the *ordered 5-tuple* $(4, 7, 8, 17, 20)$.

We can think of some vectors as having local coordinate systems that are *offset* from the origin. In computer graphics the distinction is that coordinate systems are measured in world coordinates and vectors are terminal points that are local to that coordinate system (see figure 3.6).

Example   Example vectors in $\mathbb{R}$ in figure 3.5.

$$\mathbf{a} = \mathbf{b} = \mathbf{c},$$

$$\mathbf{d} \neq \mathbf{e} \neq \mathbf{a}.$$

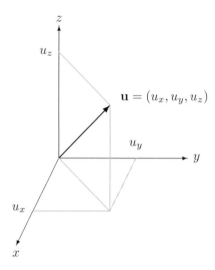

Fig. 3.4   A 3D vector with components u_x, u_y, u_z.

3.8.1.2   *Two Interesting Properties of Vectors in $\mathbb{R}^3$*

Vectors in $\mathbb{R}^3$ are represented here by a **bolded** letter. Let $\mathbf{u} = (u_x, u_y, u_z)$
and $\mathbf{v} = (v_x, v_y, v_z)$ (two vectors). An important operation is the *dot product* (used below to get the angle between two vectors):

$$\mathbf{u} \cdot \mathbf{v} = u_x v_x + u_y v_y + u_z v_z. \tag{3.55}$$

The dot ($\cdot$) here means the *inner*, or dot product. This operation takes two
vectors and returns a number (not a vector).

Knowing the components we can calculate the *magnitude* (the length of the
vector) using Pythagoras' theorem as follows:

$$\|\mathbf{u}\| = \sqrt{u_x u_x + u_y u_y + u_z u_z}. \tag{3.56}$$

Example   if $\mathbf{u} = (1, 1, 1)$ then $\|\mathbf{u}\| = \sqrt{3}$.

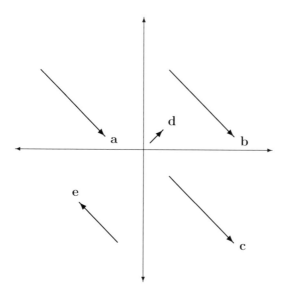

Fig. 3.5   Vector examples.

3.8.1.3   *Vectors in* $\mathbb{C}$

$V = \mathbb{C}^n$ is a *complex vector space* with dimension n. This is the set containing all column vectors with n complex numbers laid out vertically (the above examples were row vectors with components laid out horizontally). We also define a vector subspace as a nonempty set of vectors which satisfy the same conditions as the parent's vector space.

3.8.2   *Column Notation*

In $\mathbb{C}^2$ for example, the quantum mechanical notation for a *ket* can be used to represent a vector.

$$|u\rangle = \begin{bmatrix} u_1 \\ u_2 \end{bmatrix} \tag{3.57}$$

where $u_1 = a_1 + b_1 i$ and $u_2 = a_2 + b_2 i$.

$|u\rangle$ can also be represented in row form:

$$|u\rangle = (u_1, u_2) \tag{3.58}$$

Example   Column Notation.

$$|0\rangle = \begin{bmatrix} 1 \\ 0 \end{bmatrix} \text{ and, } |1\rangle = \begin{bmatrix} 0 \\ 1 \end{bmatrix}.$$

Example   Here's a more complex example:

$$|u\rangle = (1+i)|0\rangle + (2-3i)|1\rangle$$

$$= \begin{bmatrix} 1+i \\ 2-3i \end{bmatrix}.$$

3.8.3   *The Zero Vector*

The **0** vector is the vector where all entries are 0.

3.8.4   *Properties of Vectors in $\mathbb{C}^n$*

Here we take scalars $(\alpha, \beta) \in \mathbb{C}$ and vectors $(|u\rangle, |v\rangle, |w\rangle) \in \mathbb{C}^n$.

3.8.4.1   *Scalar Multiplication and Addition*

Listed here are some basic properties of complex scalar multiplication and addition.

$$\alpha|u\rangle = \begin{bmatrix} \alpha u_1 \\ \vdots \\ \alpha u_n \end{bmatrix}. \tag{3.59}$$

$$\alpha(\beta|u\rangle) = \alpha\beta|u\rangle. \tag{3.60}$$

$$\alpha(|u\rangle + |v\rangle) = \alpha|u\rangle + \alpha|v\rangle. \qquad \textit{Associative law, scalar multiplication} \quad (3.61)$$

$$(\alpha + \beta)|u\rangle = \alpha|u\rangle + \beta|u\rangle. \qquad \textit{Distributive law, scalar addition} \qquad (3.62)$$

$$\alpha(|u\rangle + |v\rangle) = \alpha|u\rangle + \alpha|v\rangle. \qquad\qquad\qquad\qquad\qquad\qquad (3.63)$$

3.8.4.2 *Vector Addition*

A sum of vectors can be represented by:

$$|u\rangle + |v\rangle = \begin{bmatrix} u_1 + v_1 \\ \vdots \\ u_n + v_n \end{bmatrix}. \qquad (3.64)$$

This sum has the following properties:

$$|u\rangle + |v\rangle = |v\rangle + |u\rangle. \qquad\qquad\qquad \textit{Commutative} \quad (3.65)$$

$$(|u\rangle + |v\rangle) + |w\rangle = |u\rangle + (|v\rangle + |w\rangle). \qquad \textit{Associative} \quad (3.66)$$

$$|u\rangle + \mathbf{0} = |u\rangle. \qquad\qquad\qquad\qquad\qquad\qquad (3.67)$$

For every $|u\rangle \in \mathbb{C}^n$ there is a corresponding unique vector $-|u\rangle$ such that:

$$|u\rangle + (-|u\rangle) = \mathbf{0}. \qquad (3.68)$$

3.8.5 *The Dual Vector*

The *dual vector* $\langle u|$ corresponding to a ket vector $|u\rangle$ is obtained by transposing the corresponding column vector and conjugating its components. This is called, in quantum mechanics, a *bra* and we have:

$$\langle u| = |u\rangle^\dagger = [u_1^*, u_2^*, \dots, u_n^*]. \qquad (3.69)$$

The dagger symbol, † is called the adjoint and is introduced in section 3.8.19.

Example   The dual of $|0\rangle$.

$$\langle 0| = |0\rangle^\dagger = [1^*, 0^*].$$

Example   Given the vector $|u\rangle$ where:

$$|u\rangle = \begin{bmatrix} 1 - i \\ 1 + i \end{bmatrix}.$$

The dual of $|u\rangle$ is:

$$\langle u| = [(1 - i)^*, (1 + i)^*]$$
$$= [(1 + i), (1 - i)]$$

3.8.6   *Linear Combinations*

A vector $|u\rangle$ is a *linear combination* of vectors $|v_1\rangle, |v_2\rangle, \ldots, |v_n\rangle$ if $|u\rangle$ can be expressed by:

$$|u\rangle = \alpha_1|v_1\rangle + \alpha_2|v_2\rangle + \ldots + \alpha_n|v_n\rangle \tag{3.70}$$

where scalars $\alpha_1, \alpha_2, \ldots, \alpha_n$ are complex numbers.

We can represent a linear combination as:

$$|u\rangle = \sum_{i=1}^{n} \alpha_i|v_i\rangle. \tag{3.71}$$

3.8.7 *Linear Independence*

A set of nonzero vectors $|v_1\rangle, ..., |v_n\rangle$ is *linearly independent* if:

$$\sum_{i=1}^{n} a_i |v_i\rangle = \mathbf{0} \text{ iff } a_1 = ... = a_n = 0. \tag{3.72}$$

Example Linear dependence. The row vectors (bras) $[1, -1], [1, 2]$, and $[2, 2]$ are linearly dependent because:

$$[1, -1] + [1, 2] - [2, 1] = [0, 0]$$

i.e there is a linear combination with $a_1 = 1, a_2 = 1, a_3 = -1$ (other than the zero condition above) that evaluates to 0; So they are not linearly independent.

3.8.8 *Spanning Set*

A *spanning set* is a set of vectors $|v_1\rangle, ..., |v_n\rangle$ for V in terms of which every vector in V can be written as a linear combination.

Example Vectors $\mathbf{u} = [1, 0, 0]$, $\mathbf{v} = [0, 1, 0]$, and $\mathbf{w} = [0, 0, 1]$ span $\mathbb{R}^3$ because all vectors $[x, y, z]$ in $\mathbb{R}^3$ can be written as a linear combination of $\mathbf{u}$, $\mathbf{v}$, and $\mathbf{w}$ like the following:

$$[x, y, z] = x\mathbf{u} + y\mathbf{v} + z\mathbf{w}.$$

3.8.9 *Basis*

A *basis* is any set of vectors that are a spanning set and are linearly independent.

Most of the time with quantum computing we'll use a standard basis, called the *computational basis*. This is also called an orthonormal basis (see section 3.8.15). In $\mathbb{C}^2$ we can use $|0\rangle$ and $|1\rangle$ for the basis. In $\mathbb{C}^4$ we can use $|00\rangle, |01\rangle, |10\rangle$, and $|11\rangle$ for a basis (the tensor product is needed to understand this — see section 3.8.29).

3.8.10   *Probability Theory*

An event x_i has a probability p_i of occurring which lies in the range:

$$0 \leq p_i \leq 1 \tag{3.73}$$

A probability of $p_i = 0$ means x_i is impossible and a probability of $p_i = 1$ means x_i is certain.

Say there are n events and the *xth* event happens n_x times and also we have $\sum_{x=1} n_x = n$. The *xth* event will occur with the following probability:

$$p_x = \frac{n_x}{n} \tag{3.74}$$

This implies the probabilities all sum to 1.

We call the average of the probabilities the *expectation value* which is defined as:

$$\langle x \rangle = \sum_{x=1} x p_x = n \tag{3.75}$$

Example A group of salesmen make a number of sales.

Sales	Salesmen
10	6
20	12
30	6

The number of salesmen is $n = 6 + 12 + 6 = 24$.

The probability of a salesman making 20 sales is:

$$p_2 = \frac{n_2}{n} = \frac{12}{24} = 0.5.$$

The expectation value $\langle x \rangle$ is:

$$\langle x \rangle = \sum_{x=1} x p_x = 10(0.25) + 20(0.5) + 30(0.25) = 20.$$

3.8.11 *Probability Amplitudes*

We can write a vector as a combination of basis vectors. In quantum mechanics we use a *state vector* $|\Psi\rangle$. The state vector in $\mathbb{C}^2$ is often written as $|\Psi\rangle = \alpha|0\rangle + \beta|1\rangle$.

The scalars (e.g. α and β in $\alpha|0\rangle + \beta|1\rangle$) associated with our basis vectors are called *probability amplitudes*, because in quantum mechanics they give the probabilities of projecting the state into a basis state, $|0\rangle$ or $|1\rangle$, when the appropriate measurement is performed (see chapter 4).

To fit with the probability interpretation the square of the absolute values of the probability amplitudes must sum to 1:

$$|\alpha|^2 + |\beta|^2 = 1. \tag{3.76}$$

Example Determine the probabilities of measuring a $|0\rangle$ or a $|1\rangle$ for $\sqrt{\frac{1}{3}}|0\rangle + \sqrt{\frac{2}{3}}|1\rangle$.

First, check if the probabilities sum to 1.

$$\left|\sqrt{\frac{1}{3}}\right|^2 + \left|\sqrt{\frac{2}{3}}\right|^2 = \frac{1}{3} + \frac{2}{3}$$

$$= 1.$$

They do sum to 1 so convert to percentages.

$$\frac{1}{3}(100) = 33.\dot{3} \text{ and } \frac{2}{3}(100) = 66.\dot{6} .$$

So this give us a 33.$\dot{3}$% chance of measuring a $|0\rangle$ and a 66.$\dot{6}$% chance of measuring a $|1\rangle$.

3.8.12 The Inner Product

We've already met the inner (or dot) product in $\mathbb{R}^2$ in section 3.8. The inner product in quantum computing is defined in terms of $\mathbb{C}^n$, but it's helpful to think of what the inner product gives us in $\mathbb{R}^2$, which is the angle between two vectors. The dot product in $\mathbb{R}^2$ is shown here (also see figure 3.6):

$$\mathbf{u} \cdot \mathbf{v} = \|\mathbf{u}\|\|\mathbf{v}\| \cos\theta \tag{3.77}$$

and rearranging we get:

$$\theta = \cos^{-1}\left(\frac{\mathbf{u} \cdot \mathbf{v}}{\|\mathbf{u}\|\|\mathbf{v}\|}\right). \tag{3.78}$$

Now we'll look at the inner product in $\mathbb{C}^n$, which is defined in terms of a dual. An inner product in $\mathbb{C}^n$ combines two vectors and produces a complex number. So given,

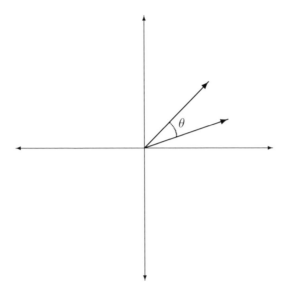

Fig. 3.6   The dot product.

$$|u\rangle = \begin{bmatrix} \alpha_1 \\ \vdots \\ \alpha_n \end{bmatrix}$$

and,

$$|v\rangle = \begin{bmatrix} \beta_1 \\ \vdots \\ \beta_n \end{bmatrix}$$

we can calculate the inner product:

$$[\alpha_1^*, ..., \alpha_n^*] \begin{bmatrix} \beta_1 \\ \vdots \\ \beta_n \end{bmatrix} = \langle u| \times |v\rangle \tag{3.79}$$

$$= \langle u|v\rangle. \tag{3.80}$$

An inner product can also be represented in the following format:

$$(|u\rangle, |v\rangle) = \langle u|v\rangle. \tag{3.81}$$

So for $\mathbb{C}^2$ the following are equivalent:

$$\left(\begin{bmatrix} u_1 \\ u_2 \end{bmatrix}, \begin{bmatrix} v_1 \\ v_2 \end{bmatrix}\right) = (|u\rangle, |v\rangle) = \langle u| \times |v\rangle = \langle u|v\rangle = [u_1^*, u_2^*] \begin{bmatrix} v_1 \\ v_2 \end{bmatrix} = u_1^* v_1 + u_2^* v_2.$$

Example Using the inner product notation from above we can extract a probability amplitude if we use one of the basis vectors as the original vector's dual:

$$\langle 0|(\alpha|0\rangle + \beta|1\rangle) = [1^*, 0^*] \begin{bmatrix} \alpha \\ \beta \end{bmatrix} = \alpha$$

or using dot product notation,

$$\langle 0|(\alpha|0\rangle + \beta|1\rangle) = \begin{bmatrix} 1 \\ 0 \end{bmatrix} \bullet \begin{bmatrix} \alpha \\ \beta \end{bmatrix} = \alpha.$$

This is called *bra-ket* notation. Hilbert space is the vector space for complex inner products.

Properties:

$$\langle u|v\rangle = \langle v|u\rangle^*. \tag{3.82}$$

$$\langle u|\alpha v\rangle = \langle \alpha^* u|v\rangle = \alpha\langle u|v\rangle. \tag{3.83}$$

$$\langle u|v + w\rangle = \langle u|v\rangle + \langle u|w\rangle. \tag{3.84}$$

$$\forall |u\rangle [\mathbb{R} \ni \langle u|u\rangle \geq 0]. \tag{3.85}$$

If $\langle u|u\rangle = 0$ then $|u\rangle = 0$. $\tag{3.86}$

$$|\langle u|v\rangle|^2 \leq \langle u|u\rangle\langle v|v\rangle. \qquad \textit{The Cauchy–Schwartz inequality} \tag{3.87}$$

3.8.13 *Orthogonality*

Orthogonal vectors can be thought of as being "perpendicular to each other"; two vectors are orthogonal iff:

$$\langle u|v \rangle = 0. \tag{3.88}$$

Example The vectors:

$$|u\rangle = \begin{bmatrix} 1 \\ 0 \end{bmatrix} \text{ and } |v\rangle = \begin{bmatrix} 0 \\ 1 \end{bmatrix}$$

are orthogonal because:

$$[1^*, 0^*] \begin{bmatrix} 0 \\ 1 \end{bmatrix} = 0.$$

Example The vectors:

$$|u\rangle = \begin{bmatrix} 1 \\ 1 \end{bmatrix} \text{ and } |v\rangle = \begin{bmatrix} 1 \\ -1 \end{bmatrix}$$

are orthogonal because:

$$\begin{aligned} \langle u|v \rangle &= ((1,1),(1,-1)) \\ &= 1 \times 1 + 1 \times (-1) \\ &= 0. \end{aligned}$$

Example The vectors:

$$|u\rangle = \begin{bmatrix} 1+i \\ 2-2i \end{bmatrix} \text{ and } |v\rangle = \begin{bmatrix} -i \\ \frac{1}{2} \end{bmatrix}$$

are orthogonal because:

$$[1-i, 2+2i] \begin{bmatrix} -i \\ \frac{1}{2} \end{bmatrix} = 0.$$

3.8.14   *The Unit Vector*

A vector's norm is:

$$\||u\rangle\| = \sqrt{\langle u|u \rangle} \ . \tag{3.89}$$

A *unit* vector is a vector where its norm is equal to 1.

$$\||u\rangle\| = 1. \tag{3.90}$$

If we want to make an arbitrary vector a unit vector, which here is represented by the ˆ (hat) symbol, we must *normalise* it by dividing by the norm:

$$|\hat{u}\rangle = \frac{|u\rangle}{\||u\rangle\|} \ . \tag{3.91}$$

Example   Normalise $|u\rangle = \begin{bmatrix} 1 \\ 1 \end{bmatrix}$ $(= |0\rangle + |1\rangle)$.

First we find the norm:

$$\||u\rangle\| = \sqrt{[1^*, 1^*]\begin{bmatrix} 1 \\ 1 \end{bmatrix}}$$
$$= \sqrt{2}\ .$$

Now we normalise $\||u\rangle\|$ to get:

$$|\hat{u}\rangle = \frac{\begin{bmatrix} 1 \\ 1 \end{bmatrix}}{\sqrt{2}}$$
$$= \frac{1}{\sqrt{2}}\begin{bmatrix} 1 \\ 1 \end{bmatrix}$$
$$= \begin{bmatrix} \frac{1}{\sqrt{2}} \\ \frac{1}{\sqrt{2}} \end{bmatrix}$$
$$= \frac{1}{\sqrt{2}}|0\rangle + \frac{1}{\sqrt{2}}|1\rangle.$$

3.8.15   *Bases for $\mathbb{C}^n$*

$\mathbb{C}^n$ has a standard basis:

$$\begin{bmatrix} 1 \\ 0 \\ \dots \\ 0 \end{bmatrix}, \begin{bmatrix} 0 \\ 1 \\ \dots \\ 0 \end{bmatrix}, \dots, \begin{bmatrix} 0 \\ 0 \\ \dots \\ 1 \end{bmatrix}. \tag{3.92}$$

This is written as $|0\rangle, |1\rangle, \dots, |n-1\rangle$. Any other vector in the same space can be expanded in terms of $|0\rangle, |1\rangle, ..., |n-1\rangle$. The basis is called *orthonormal* because the vectors are of unit length and mutually orthogonal. There are other orthonormal bases in $\mathbb{C}^n$ and, in quantum computing, it is sometimes convenient to switch between bases.

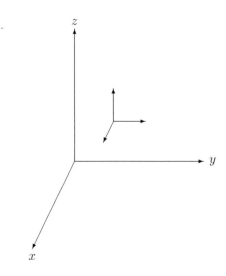

Fig. 3.7   A local coordinate system.

Example   Orthonormal bases $|0\rangle, |1\rangle$ and $\frac{1}{\sqrt{2}}(|0\rangle + |1\rangle)$, $\frac{1}{\sqrt{2}}(|0\rangle - |1\rangle)$ are often used for quantum computing.

It is useful at this point to consider what an orthonormal basis is in $\mathbb{R}^3$. The "ortho" part of orthonormal stands for orthogonal, which means the vectors are perpendicular to each other, for example the 3D axes (x, y, z) are orthogonal. The "normal" part refers to normalised (unit) vectors. We can use an orthonormal basis in $\mathbb{R}^3$ to separate world coordinates from a local coordinate system in 3D computer graphics. In this system we define the position of the local coordinate system in world coordinates and then we can define the positions of individual objects in terms of the local coordinate system. In this way we can transform the local coordinates system, together with everything in it, while leaving the world coordinates system intact. In figure 3.7 the local coordinate system forms an orthonormal basis.

3.8.16 *The Gram Schmidt Method*

Suppose $|u_1\rangle, \ldots, |u_n\rangle$ is a basis (any basis will do) for vector space V that has an inner product. Suppose this basis is not orthonormal.

The *Gram Schmidt method* can be used to produce an orthonormal basis set $|v_1\rangle, \ldots, |v_n\rangle$ for V by,

$$|v_1\rangle = \frac{|u_1\rangle}{\||u_1\rangle\|} \tag{3.93}$$

and given $|v_{k+1}\rangle$ for $1 \le k < n - 1$:

$$|v_{k+1}\rangle = \frac{|u_{k+1}\rangle - \sum_{i=1}^{k}\langle v_i|u_{k+1}\rangle|v_i\rangle}{\||u_{k+1}\rangle - \sum_{i=1}^{k}\langle v_i|u_{k+1}\rangle|v_i\rangle\|} . \tag{3.94}$$

Example Given the following vectors in $\mathbb{C}^3$: $|u_1\rangle = (i, i, i), |u_2\rangle = (0, i, i)$, and $|u_3\rangle = (0, 0, i)$ find an orthonormal basis $|v_1\rangle, |v_2\rangle, |v_3\rangle$.

$$
\begin{aligned}
|v_1\rangle &= \frac{|u_1\rangle}{\||u_1\rangle\|} \\
&= \frac{(i, i, i)}{\sqrt{3}} \\
&= \left(\frac{i}{\sqrt{3}}, \frac{i}{\sqrt{3}}, \frac{i}{\sqrt{3}}\right) \\
|v_2\rangle &= \frac{|u_2\rangle - \langle v_1|u_2\rangle|v_1\rangle}{\||u_2\rangle - \langle v_1|u_2\rangle|v_1\rangle\|} \\
&= \left(-\frac{2i}{\sqrt{6}}, \frac{i}{\sqrt{6}}, \frac{i}{\sqrt{6}}\right) \\
|v_3\rangle &= \frac{|u_3\rangle - \langle v_1|u_3\rangle|v_1\rangle - \langle v_2|u_3\rangle|v_2\rangle}{\||u_3\rangle - \langle v_1|u_3\rangle|v_1\rangle - \langle v_2|u_3\rangle|v_2\rangle\|} \\
&= \left(0, -\frac{i}{2}, \frac{i}{2}\right) .
\end{aligned}
$$

3.8.17 *Linear Operators*

A *linear operator* $A : V \to W$ where V and W are complex vector spaces is defined as:

$$A(\alpha|u\rangle + \beta|v\rangle) = \alpha(A(|u\rangle)) + \beta(A|v\rangle))). \tag{3.95}$$

With dimensions n for V and m for W the linear operator can be represented by an $m \times n$ matrix.

Example Given a linear operator A, apply it to $\sqrt{\frac{1}{3}}|0\rangle + \sqrt{\frac{2}{3}}|1\rangle$:

$$A = \begin{bmatrix} 0 & 1 \\ 1 & 0 \end{bmatrix}, |0\rangle = \begin{bmatrix} 1 \\ 0 \end{bmatrix}, |1\rangle = \begin{bmatrix} 0 \\ 1 \end{bmatrix}.$$

$$A\left(\sqrt{\frac{1}{3}}|0\rangle + \sqrt{\frac{2}{3}}|1\rangle\right) = \begin{bmatrix} 0 & 1 \\ 1 & 0 \end{bmatrix}\left(\sqrt{\frac{1}{3}}|0\rangle + \sqrt{\frac{2}{3}}|1\rangle\right)$$

$$= \begin{bmatrix} 0 & 1 \\ 1 & 0 \end{bmatrix}\begin{bmatrix} \sqrt{\frac{1}{3}} \\ \sqrt{\frac{2}{3}} \end{bmatrix}$$

$$= \begin{bmatrix} \sqrt{\frac{2}{3}} \\ \sqrt{\frac{1}{3}} \end{bmatrix}$$

$$= \sqrt{\frac{2}{3}}|0\rangle + \sqrt{\frac{1}{3}}|1\rangle.$$

Properties:

$\langle u|A|v\rangle$ is the inner product of $\langle u|$ and $A|v\rangle$. $\tag{3.96}$

3.8.18   *Outer Products and Projectors*

We define an *outer product*, $|u\rangle\langle v|$ as a linear operator A which does the following:

$$(|u\rangle\langle v|)(|w\rangle) = |u\rangle\langle v|w\rangle = \langle v|w\rangle|u\rangle. \tag{3.97}$$

This can be read as,

1. The result of the linear operator $|u\rangle\langle v|$ acting on $|w\rangle$

or,

2. The result of multiplying $|u\rangle$ by $\langle v|w\rangle$.

In terms of matrices, $|u\rangle\langle v|$ can be represented by:

$$\begin{bmatrix} u_1 \\ u_2 \\ \vdots \end{bmatrix} \begin{bmatrix} v_1^* & v_2^* & \cdots \end{bmatrix} = \begin{bmatrix} u_1 v_1^* & u_1 v_2^* & \cdots & u_1 v_n^* \\ u_2 v_1^* & u_2 v_2^* & \cdots & u_2 v_n^* \\ \vdots & & \ddots & \\ u_n v_1^* & u_n v_2^* & \cdots & u_n v_n^* \end{bmatrix}. \tag{3.98}$$

Example Take $|\Psi\rangle = \alpha|0\rangle + \beta|1\rangle$ then:

$$\begin{aligned} |1\rangle\langle 1|\Psi\rangle &= |1\rangle\langle 1|(\alpha|0\rangle + \beta|1\rangle) \\ &= |1\rangle\beta \\ &= \beta|1\rangle, \\ |0\rangle\langle 1|\Psi\rangle &= |0\rangle\langle 1|(\alpha|0\rangle + \beta|1\rangle) \\ &= |0\rangle\beta \\ &= \beta|0\rangle, \\ |1\rangle\langle 0|\Psi\rangle &= |1\rangle\langle 0|(\alpha|0\rangle + \beta|1\rangle) \\ &= |1\rangle\alpha \\ &= \alpha|1\rangle, \\ |0\rangle\langle 0|\Psi\rangle &= |0\rangle\langle 0|(\alpha|0\rangle + \beta|1\rangle) \\ &= |0\rangle\alpha \\ &= \alpha|0\rangle. \end{aligned}$$

In the chapters ahead we will use *projectors* to deal with quantum measurements. Say we have a vector space $V = \{|00\rangle, |01\rangle, |10\rangle, |11\rangle\}$. A projector P on to the subspace $V_s = \{|00\rangle, |01\rangle\}$ behaves as follows:

$$P(\alpha_{00}|00\rangle + \alpha_{01}|01\rangle + \alpha_{10}|10\rangle + \alpha_{11}|11\rangle) = \alpha_{00}|00\rangle + \alpha_{01}|01\rangle.$$

P projects any vector in V onto V_s (components not in V_s are discarded). We can represent projectors with outer product notation. Given a subspace which is spanned by orthonormal vectors, $\{|u_1\rangle, |u_2\rangle \ldots, |u_n\rangle\}$, a projection onto this subspace can be represented by a summation of outer products:

$$P = \sum_{i=1}^{n} |u_i\rangle\langle u_i| \tag{3.99}$$

$$= |u_1\rangle\langle u_1| + |u_2\rangle\langle u_2| + \ldots + |u_n\rangle\langle u_n|. \tag{3.100}$$

So, we can replace the projector notation P with the explicit outer product notation:

$$(|00\rangle\langle 00| + |01\rangle\langle 01|)(\alpha_{00}|00\rangle + \alpha_{01}|01\rangle + \alpha_{10}|10\rangle + \alpha_{11}|11\rangle) = \alpha_{00}|00\rangle + \alpha_{01}|01\rangle.$$

We can also represent a matrix (an operator for example) using outer product notation, as shown in the next example.

Example Representing operators X and Z. These two matrices are defined below, but it turns out they are quite handy for quantum computing. We'll be using them frequently in the chapters ahead:

$$|0\rangle\langle 1| = \begin{bmatrix} 1 \\ 0 \end{bmatrix} \begin{bmatrix} 0^* & 1^* \end{bmatrix}$$

$$= \begin{bmatrix} 0 & 1 \\ 0 & 0 \end{bmatrix},$$

$$|1\rangle\langle 1| = \begin{bmatrix} 0 \\ 1 \end{bmatrix} \begin{bmatrix} 0^* & 1^* \end{bmatrix}$$

$$= \begin{bmatrix} 0 & 0 \\ 0 & 1 \end{bmatrix},$$

$$|0\rangle\langle 0| = \begin{bmatrix} 1 \\ 0 \end{bmatrix} \begin{bmatrix} 1^* & 0^* \end{bmatrix}$$

$$= \begin{bmatrix} 1 & 0 \\ 0 & 0 \end{bmatrix},$$

$$|1\rangle\langle 0| = \begin{bmatrix} 0 \\ 1 \end{bmatrix} \begin{bmatrix} 1^* & 0^* \end{bmatrix}$$

$$= \begin{bmatrix} 0 & 0 \\ 1 & 0 \end{bmatrix}.$$

$$X = \begin{bmatrix} 0 & 1 \\ 1 & 0 \end{bmatrix}$$

$$= |0\rangle\langle 1| + |1\rangle\langle 0|,$$

$$Z = \begin{bmatrix} 1 & 0 \\ 0 & -1 \end{bmatrix}$$

$$= |0\rangle\langle 0| - |1\rangle\langle 1|.$$

Properties:

$$\sum_i |i\rangle\langle i| = I \text{ for any orthonormal basis } \{|i\rangle\}.$$

This is the *Completeness relation*   (3.101)

Each component $|u\rangle\langle u|$ of P is hermitian

and P itself is hermitian (see 3.8.24).   (3.102)

$P^\dagger = P$ See the next section on the adjoint.   (3.103)

$P^2 = P.$ (3.104)

$Q = I - P$ is called the *orthogonal complement*   (3.105)

Later we'll look at quantum measurements, and we'll use M_m to represent a measurement. If we use a projector (i.e. $M_m = P$) for measurement then the probability of measuring m is:

$$p_m = \langle\Psi| M_m^\dagger M_m |\Psi\rangle.$$

By 3.103 and 3.104 we can say that this is equivalent to:

$$p_m = \langle\Psi| M_m |\Psi\rangle.$$

3.8.19   *The Adjoint*

The *adjoint* $A^\dagger$ is the matrix obtained from A by conjugating all the elements of A (to get A^*) and then forming the transpose:

$$A^\dagger = (A^*)^T.$$ (3.106)

Example   An adjoint.

$$\begin{bmatrix} 1+i & 1-i \\ -1 & 1 \end{bmatrix}^\dagger = \begin{bmatrix} 1-i & -1 \\ 1+i & 1 \end{bmatrix}.$$

Properties:

$$\langle u|Av\rangle = \langle A^\dagger u|v\rangle. \tag{3.107}$$

$$(AB)^\dagger = B^\dagger A^\dagger. \tag{3.108}$$

$$(A^\dagger)^\dagger = A. \tag{3.109}$$

$$(|u\rangle)^\dagger = \langle u|. \tag{3.110}$$

$$(A|u\rangle)^\dagger = \langle u|A^\dagger \text{ but not } A|u\rangle = \langle u|A^\dagger. \tag{3.111}$$

$$(\alpha A + \beta B)^\dagger = \alpha^* A^\dagger + \beta^* B^\dagger. \tag{3.112}$$

Example   Example of $\langle u|Av\rangle = \langle A^\dagger u|v\rangle$. Given,

$$|u\rangle = \begin{bmatrix} 1 \\ i \end{bmatrix}, |v\rangle = \begin{bmatrix} 1 \\ 1 \end{bmatrix} \text{ and } A = \begin{bmatrix} 1+i & 1-i \\ -1 & 1 \end{bmatrix}.$$

$$A|v\rangle = \begin{bmatrix} 1+i & 1-i \\ -1 & 1 \end{bmatrix} \begin{bmatrix} 1 \\ 1 \end{bmatrix}$$

$$= \begin{bmatrix} 2 \\ 0 \end{bmatrix},$$

$$\langle u|Av\rangle = 2,$$

$$A^\dagger|u\rangle = \begin{bmatrix} 1-i & -1 \\ 1+i & 1 \end{bmatrix} \begin{bmatrix} 1 \\ i \end{bmatrix}$$

$$= \begin{bmatrix} 1-2i \\ 1+2i \end{bmatrix},$$

$$\langle A^\dagger u|v\rangle = \begin{bmatrix} 1+2i & 1-2i \end{bmatrix} \begin{bmatrix} 1 \\ 1 \end{bmatrix}$$

$$= 2.$$

3.8.20 *Eigenvalues and Eigenvectors*

The complex number λ is an *eigenvalue* of a linear operator A if there exists vector $|u\rangle$ such that:

$$A|u\rangle = \lambda|u\rangle \qquad (3.113)$$

where $|u\rangle$ is called an *eigenvector* of A.

Eigenvalues of A can be found by using the following equation, called the *characteristic equation* of A:

$$c(\lambda) = \det(A - \lambda I) = 0. \qquad (3.114)$$

This comes from noting that:

$$A|u\rangle = \lambda|u\rangle \Leftrightarrow (A - \lambda I)|u\rangle = 0 \Leftrightarrow A - \lambda I \text{ is singular } \Leftrightarrow \det(A - \lambda I) = 0.$$

Solving the characteristic equation gives us the *characteristic polynomial* of A. We can then solve the characteristic polynomial to find all the eigenvalues for A. If A is an $n \times n$ matrix, there will be n eigenvalues (but some may be the same as others).

Properties:

A's i^{th} eigenvalue λ_i has eigenvector $|u_i\rangle$ iff $A|u_i\rangle = \lambda_i|u_i\rangle$ $\qquad (3.115)$

An *eigenspace* for λ_i is the set of eigenvectors that satisfies

$\qquad A|u_j\rangle = \lambda_i|u_j\rangle$, here j is the index for eigenvectors of λ_i. $\qquad (3.116)$

An eigenspace is *degenerate* when it has dimension > 1

$\qquad$ i.e. more than one eigenvector. $\qquad (3.117)$

Note: The eigenvectors that match different eigenvalues are linearly independent, which means we can have an orthonormal set of eigenvectors for an operator A.

An example is on the next page.

Example Eigenvalues and eigenvectors of X.

$$X = \begin{bmatrix} 0 & 1 \\ 1 & 0 \end{bmatrix}.$$

$$\det(X - \lambda I) = \begin{bmatrix} -\lambda & 1 \\ 1 & -\lambda \end{bmatrix}$$
$$= \lambda^2 - 1.$$

This is the characteristic polynomial. The two solutions to $\lambda^2 - 1 = 0$ are $\lambda = -1$ and $\lambda = +1$. If we use the eigenvalue of $\lambda = -1$ to determine the corresponding eigenvector $|\lambda_{-1}\rangle$ of X we get:

$$X|\lambda_{-1}\rangle = -1|\lambda_{-1}\rangle$$
$$\begin{bmatrix} 0 & 1 \\ 1 & 0 \end{bmatrix} \begin{bmatrix} \alpha \\ \beta \end{bmatrix} = \begin{bmatrix} -\alpha \\ -\beta \end{bmatrix}$$
$$\begin{bmatrix} \beta \\ \alpha \end{bmatrix} = \begin{bmatrix} -\alpha \\ -\beta \end{bmatrix}.$$

We get $\alpha = -\beta$, so after normalisation our eigenvector is:

$$|\lambda_{-1}\rangle = \frac{1}{\sqrt{2}}|0\rangle - \frac{1}{\sqrt{2}}|1\rangle.$$

Notice that we've used the eigenvalue $\lambda = -1$ to label the eigenvector $|\lambda_{-1}\rangle$.

3.8.21 *Trace*

The *trace* of A is the sum of its eigenvalues, or:

$$\mathrm{tr}(A) = \sum_{i=1}^{n} a_{ii} \tag{3.118}$$

i.e. the sum of its diagonal entries.

Example   Trace of X and I

$$X = \begin{bmatrix} 0 & 1 \\ 1 & 0 \end{bmatrix}.$$

$$I = \begin{bmatrix} 1 & 0 \\ 0 & 1 \end{bmatrix}.$$

$\mathrm{tr}(X) = 0 + 0 = 0$ or the sum of the eigenvalues $1 + (-1) = 0$. For I we have $\mathrm{tr}(I) = 2$.

Further properties are listed below, where U is unitary. See section 3.8.23 for an explanation of unitary operators:

$$\mathrm{tr}(A + B) = \mathrm{tr}(A) + \mathrm{tr}(B). \tag{3.119}$$

$$\mathrm{tr}(\alpha(A + B)) = \alpha\mathrm{tr}(A) + \alpha\mathrm{tr}(B). \tag{3.120}$$

$$\mathrm{tr}(AB) = \mathrm{tr}(BA). \tag{3.121}$$

$$\mathrm{tr}(|u\rangle\langle v|) = \langle u|v\rangle. \tag{3.122}$$

$$\mathrm{tr}(\alpha A) = \alpha\mathrm{tr}(A). \tag{3.123}$$

$$\mathrm{tr}(UAU^{\dagger}) = \mathrm{tr}(A). \text{ } \textit{Similarity transform for U} \tag{3.124}$$

$$\mathrm{tr}(U^{\dagger}AU) = \mathrm{tr}(A). \tag{3.125}$$

$$\mathrm{tr}(A|u\rangle\langle u|) = \langle u|A|u\rangle \text{ if } |u\rangle \text{ is unitary.} \tag{3.126}$$

For unit norm $|u\rangle$:

$$\mathrm{tr}(|u\rangle\langle u|) = \mathrm{tr}(|u\rangle\langle u||u\rangle\langle u|) \tag{3.127}$$

$$= \langle u||u\rangle\langle u||u\rangle \tag{3.128}$$

$$= \langle u|u\rangle\langle u|u\rangle \tag{3.129}$$

$$= ||u\rangle|^4 \tag{3.130}$$

$$= 1. \tag{3.131}$$

As stated, the trace of A is the sum of its eigenvalues. We can also say that:

$$U^\dagger A U = \begin{bmatrix} \lambda_1 & & \\ & \ddots & \\ & & \lambda_n \end{bmatrix}.$$

which also has a trace which is the sum of the eigenvalues of A as $\mathrm{tr}(U^\dagger A U) = \mathrm{tr}(A)$ (see section 3.8.25 below).

3.8.22 *Normal Operators*

A *normal* operator satisfies the following condition:

$$AA^\dagger = A^\dagger A. \tag{3.132}$$

The class of normal operators has a number of subsets. In the following sections we'll look at some of the important normal operators. These include *unitary*, *hermitian*, and *positive* operators. The relationships between these operators is shown in figure 3.8.

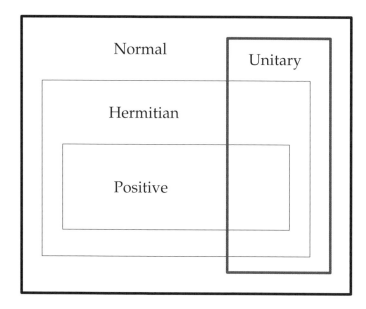

Fig. 3.8 Relationships between operators.

3.8.23   *Unitary Operators*

Matrix U is unitary (unitary operators are usually represented by U) if:

$$U^{-1} = U^\dagger \tag{3.133}$$

or,

$$UU^\dagger = U^\dagger U = I. \tag{3.134}$$

Unitary matrices preserve norm:

$$\|U|u\rangle\| = \||u\rangle\| \ \forall \ |u\rangle. \tag{3.135}$$

There are some particularly important operators called the *Pauli operators*. We've seen some of them already, they are referred to by the letters I, X, Y, and Z. In some texts X, Y, and Z are referred to by another notation, where $\sigma_1 = \sigma_X = X, \sigma_2 = \sigma_Y = Y$, and $\sigma_3 = \sigma_Z = Z$. The Pauli operators are defined as:

$$I = \begin{bmatrix} 1 & 0 \\ 0 & 1 \end{bmatrix}, \tag{3.136}$$

$$X = \begin{bmatrix} 0 & 1 \\ 1 & 0 \end{bmatrix}, \tag{3.137}$$

$$Y = \begin{bmatrix} 0 & -i \\ i & 0 \end{bmatrix}, \tag{3.138}$$

$$Z = \begin{bmatrix} 1 & 0 \\ 0 & -1 \end{bmatrix}. \tag{3.139}$$

Example $I, X, Y,$ and Z are unitary because:

$$II^\dagger = I^2 = \begin{bmatrix} 1 & 0 \\ 0 & 1 \end{bmatrix} \begin{bmatrix} 1 & 0 \\ 0 & 1 \end{bmatrix} = \begin{bmatrix} 1 & 0 \\ 0 & 1 \end{bmatrix}.$$

$$XX^\dagger = X^2 = \begin{bmatrix} 0 & 1 \\ 1 & 0 \end{bmatrix} \begin{bmatrix} 0 & 1 \\ 1 & 0 \end{bmatrix} = \begin{bmatrix} 1 & 0 \\ 0 & 1 \end{bmatrix}.$$

$$YY^\dagger = Y^2 = \begin{bmatrix} 0 & -i \\ i & 0 \end{bmatrix} \begin{bmatrix} 0 & -i \\ i & 0 \end{bmatrix} = \begin{bmatrix} 1 & 0 \\ 0 & 1 \end{bmatrix}.$$

$$ZZ^\dagger = Z^2 = \begin{bmatrix} 1 & 0 \\ 0 & -1 \end{bmatrix} \begin{bmatrix} 1 & 0 \\ 0 & -1 \end{bmatrix} = \begin{bmatrix} 1 & 0 \\ 0 & 1 \end{bmatrix}.$$

Note: $I = I^\dagger, X = X^\dagger, Y = Y^\dagger,$ and $Z = Z^\dagger$.

Properties (of unitary operators):

$$U = \sum_j |j\rangle\langle j|. \tag{3.140}$$

$$(U|u\rangle, U|v\rangle) = \langle u|U^\dagger U|v\rangle = \langle u|v\rangle. \tag{3.141}$$

Unitary matrices are also normal. $\tag{3.142}$

Unitary matrices allow for spectral decomposition,
(see section 3.8.28). $\tag{3.143}$

Unitary matrices allow for reversal, i.e. $U^\dagger(U|u\rangle) = I|u\rangle = |u\rangle$. $\tag{3.144}$

Unitary matrices preserve inner product
$$(U|u\rangle, U|v\rangle) = (|u\rangle, |v\rangle) = \langle u|v\rangle. \tag{3.145}$$

Unitary matrices preserve norm $\|U|u\rangle\| = \||u\rangle\|$. $\tag{3.146}$

Given an orthonormal basis set $\{|u_i\rangle\}, \{U|u_i\rangle\} = \{v_i\}$
is also an orthonormal basis with $U = \sum_i |v_i\rangle\langle u_i|$. $\tag{3.147}$

Unitary matrices have eigenvalues of modulus 1. $\tag{3.148}$

3.8.24 *Hermitian and Positive Operators*

A hermitian matrix A has the property:

$$A = A^\dagger. \tag{3.149}$$

The eigenvalues of a hermitian matrix are real numbers and hermitian matrices are also normal (although not all normal matrices have real eigenvalues).

Example The matrix X is Hermitian because:

$$X = \begin{bmatrix} 0 & 1 \\ 1 & 0 \end{bmatrix} =, X^\dagger = \begin{bmatrix} 0 & 1 \\ 1 & 0 \end{bmatrix}.$$

Properties:

$A = B + iC$ can represent any operator if B and C are hermitian
with $C = 0$ if A itself is hermitian. $\qquad (3.150)$

If A is hermitian then for $|u\rangle$, $\langle u|A|u\rangle \in \mathbb{R}$. $\qquad (3.151)$

If A is hermitian then A is a *positive* operator iff for any $|u\rangle$, $\langle u|Au\rangle \in \mathbb{R}$
and $\langle u|Au\rangle \geq 0$. $\qquad (3.152)$

If A is positive it has no negative eigenvalues. $\qquad (3.153)$

3.8.25 *Diagonalisable Matrix*

An operator A is diagonalisable if:

$$A = \sum_i \lambda_i |u_i\rangle\langle u_i|. \tag{3.154}$$

The vectors $|u_i\rangle$ form an orthonormal set of eigenvectors for A, with eigenvalues of λ_i. This is the same as saying that A can be transformed to:

$$\begin{bmatrix} \lambda_1 & & \\ & \ddots & \\ & & \lambda_n \end{bmatrix}. \tag{3.155}$$

Example Representing operator X:

$$X = \begin{bmatrix} 0 & 1 \\ 1 & 0 \end{bmatrix}.$$

The two normalised eigenvectors for X are:

$$\frac{1}{\sqrt{2}}|0\rangle - \frac{1}{\sqrt{2}}|1\rangle \text{ and } \frac{1}{\sqrt{2}}|0\rangle + \frac{1}{\sqrt{2}}|1\rangle.$$

The two vectors are orthogonal (with eigenvalues -1 and $+1$), X is diagonalisable and is given by:

$$|1\rangle\langle 0| + |0\rangle\langle 1|$$

3.8.26 *The Commutator and Anti-Commutator*

Here is a set of properties for the *Commutator* and *Anti-Commutator* which relate to commutative relationships between two operators A and B.

Commutator:

$$[A, B] = AB - BA, A \text{ and } B \text{ commute } (AB = BA) \text{ if } [A, B] = 0.$$
$$(3.156)$$

Anti-commutator:

$$\{A, B\} = AB + BA. \text{ We say } A \text{ and } B \text{ anti-commute if } \{A, B\} = 0.$$
$$(3.157)$$

Example We test X and Z against the commutator.

$$[X, Z] = \begin{bmatrix} 0 & 1 \\ 1 & 0 \end{bmatrix} \begin{bmatrix} 1 & 0 \\ 0 & -1 \end{bmatrix} - \begin{bmatrix} 1 & 0 \\ 0 & -1 \end{bmatrix} \begin{bmatrix} 0 & 1 \\ 1 & 0 \end{bmatrix}$$

$$= \begin{bmatrix} 0 & -2 \\ 2 & 0 \end{bmatrix}$$

$$\neq 0.$$

So X and Z do not commute.

The *simultaneous diagonalisation theorem* says that if H_A and H_B are hermitian, $[H_A, H_B] = 0$ if $\exists$ a set of orthonormal eigenvectors for both H_A, H_B so:

$$H_A = \sum_i \lambda_i' |i\rangle\langle i| \text{ and } H_B = \sum_i \lambda_i'' |i\rangle\langle i|. \qquad (3.158)$$

i.e. they are both diagonal in a common basis.

Properties:

$$AB = \frac{[A, B] + \{A, B\}}{2} . \qquad (3.159)$$

$$[A, B]^\dagger = [A^\dagger, B^\dagger]. \qquad (3.160)$$

$$[A, B] = -[B, A]. \qquad (3.161)$$

$[H_A, H_B]$ is hermitian if H_A, H_B are hermitian. $\qquad (3.162)$

3.8.27 *Polar Decomposition*

Polar decomposition says that any linear operator A can be represented as $A = U\sqrt{A^\dagger A}$ (called the left polar decomposition) $= \sqrt{AA^\dagger}U$ (called the right polar decomposition) where U is a unitary operator.

Single value decomposition says that if a linear operator A is a square matrix (i.e. the same input and output dimension) then there exist unitaries U_A and U_B, and D a diagonal matrix with non-negative elements in $\mathbb{R}$, such that $A = U_A D U_B$.

3.8.28 *Spectral Decomposition*

A linear operator is normal ($A^\dagger A = AA^\dagger$) iff it has orthogonal eigenvectors and the normalised (orthonormal) versions $\{u_i\}$ of the eigenvectors can diagonalise the operator:

$$A = \sum_i \lambda_i |u_i\rangle\langle u_i|. \qquad (3.163)$$

Example   Spectral decomposition of X and Z.

$$Z = \begin{bmatrix} 1 & 0 \\ 0 & -1 \end{bmatrix}$$

$$= |0\rangle\langle 0| - |1\rangle\langle 1|.$$

$$X = \begin{bmatrix} 0 & 1 \\ 1 & 0 \end{bmatrix}$$

$$= |+\rangle\langle +| - |-\rangle\langle -|.$$

X has eigenvectors $|+\rangle = \frac{1}{\sqrt{2}}|0\rangle + |1\rangle$, and $|-\rangle = \frac{1}{\sqrt{2}}|0\rangle - |1\rangle$ and eigenvalues of $+1$, and -1 Then, if we expand, we get back X:

$$|+\rangle\langle +| - |-\rangle\langle -| = \frac{1}{2}\begin{bmatrix} 1 \\ 1 \end{bmatrix}[1 1] - \frac{1}{2}\begin{bmatrix} 1 \\ -1 \end{bmatrix}[1 - 1]$$

$$= \frac{1}{2}\begin{bmatrix} 1 & 1 \\ 1 & 1 \end{bmatrix} - \frac{1}{2}\begin{bmatrix} 1 & -1 \\ -1 & 1 \end{bmatrix}$$

$$= \begin{bmatrix} 0 & 1 \\ 1 & 0 \end{bmatrix}.$$

Properties:

$A = UDU^\dagger$ where U is a unitary and D is a diagonal operator.   (3.164)

If A is normal then it has a spectral decomposition of $\sum_a |a\rangle\langle a|$.   (3.165)

3.8.29   *Tensor Products*

In a *tensor product* we have a combination of two smaller vector spaces to form a larger one. The elements of the smaller vector spaces are combined whilst preserving scalar multiplication and linearity. Formally:

If $\{|u\rangle\}$ and $\{|v\rangle\}$ are bases for V and W respectively then $\{|u\rangle \otimes |v\rangle\}$ form a basis for $V \otimes W$. We can write this in the following way:

$$|u\rangle \otimes |v\rangle = |u\rangle|v\rangle = |u, v\rangle = |uv\rangle.$$ (3.166)

Example A simple tensor product.

$$|1\rangle \otimes |0\rangle = |1\rangle|0\rangle = |1,0\rangle = |10\rangle.$$

The *Kronecker product* is defined as:

$$A \otimes B = \begin{bmatrix} a & b \\ c & d \end{bmatrix} \otimes \begin{bmatrix} x & y \\ v & w \end{bmatrix} \tag{3.167}$$

$$= \begin{bmatrix} a \cdot B & b \cdot B \\ c \cdot B & d \cdot B \end{bmatrix} \tag{3.168}$$

$$= \begin{bmatrix} ax & ay & bx & by \\ av & aw & bv & bw \\ cx & cy & dx & dy \\ cv & cw & dv & dw \end{bmatrix}. \tag{3.169}$$

where A and B are linear operators.

Example A Kronecker product on Pauli matrices X and Y

$$X = \begin{bmatrix} 0 & 1 \\ 1 & 0 \end{bmatrix} \text{ and } Y = \begin{bmatrix} 0 & -i \\ i & 0 \end{bmatrix}.$$

$$X \otimes Y = \begin{bmatrix} 0 \cdot Y & 1 \cdot Y \\ 1 \cdot Y & 0 \cdot Y \end{bmatrix}$$

$$= \begin{bmatrix} 0 & 1 \\ 1 & 0 \end{bmatrix} \begin{bmatrix} 0 & -i \\ i & 0 \end{bmatrix}$$

$$= \begin{bmatrix} 0 & 0 & 0 & -i \\ 0 & 0 & 1 & 0 \\ 0 & -i & 0 & 0 \\ i & 0 & 0 & 0 \end{bmatrix}.$$

Properties:

Tensor products are related to the inner product:

$$(|uv\rangle, |u'v'\rangle) = (|u\rangle, |u'\rangle)(|v\rangle, |v'\rangle) \tag{3.170}$$

$$= \langle u|u'\rangle\langle v|v'\rangle. \tag{3.171}$$

$$|u\rangle^{\otimes k} = (|u\rangle \otimes \ldots \otimes |u\rangle)_k \ k \text{ times.} \tag{3.172}$$

$$(A \otimes B)^* = A^* \otimes B^*. \tag{3.173}$$

$$(A \otimes B)^T = A^T \otimes B^T. \tag{3.174}$$

$$(A \otimes B)^\dagger = A^\dagger \otimes B^\dagger. \tag{3.175}$$

$$|ab\rangle^\dagger = \langle ab|. \tag{3.176}$$

$$\alpha(|u, v\rangle) = |\alpha u, v\rangle = |u, \alpha v\rangle. \tag{3.177}$$

$$|u_1 + u_2, v\rangle = |u_1, v\rangle + |u_2, v\rangle. \tag{3.178}$$

$$|u, v_1 + v_2\rangle = |u, v_1\rangle + |u, v_2\rangle. \tag{3.179}$$

$$|uv\rangle \neq |vu\rangle. \tag{3.180}$$

For linear operators A and B, $A \otimes B(|uv\rangle) = A|u\rangle \otimes B|v\rangle$. $\qquad$ (3.181)

For normal operators N_A and N_B, $N_A \otimes N_B$ will be normal. $\qquad$ (3.182)

For hermitian operators H_A and H_B, $H_A \otimes H_B$ will be hermitian.

$\qquad$ (3.183)

For unitary operators U_A and U_B, $U_A \otimes U_B$ will be unitary. $\qquad$ (3.184)

For positive operators P_A and P_B, $P_A \otimes P_B$ will be positive. $\qquad$ (3.185)

3.9　Fourier Transforms

The *Fourier transform*, which is named after Jean Baptiste Joseph Fourier, 1768–1830 (figure 3.9), maps data from the *time domain* to the *frequency domain*. The *discrete Fourier transform* (DFT) is a version of the Fourier transform which, unlike the basic Fourier transform, does not involve calculus and can be directly implemented on computers but is limited to periodic functions. The Fourier transform itself is not limited to periodic functions.

Fig. 3.9   Jean Baptiste Joseph Fourier.

3.9.1   *The Fourier Series*

Representing a periodic function as a linear combination of sines and cosines is called a *Fourier series* expansion of the function. We can represent any periodic, continuous function as a linear combination of sines and cosines. In fact, just like how $|0\rangle$ and $|1\rangle$ form an orthonormal basis for quantum computing, sin and cos form an orthonormal basis for the time domain based representation of a waveform. One way to describe an orthonormal basis is:

That which you measure against.

The fourier series has the form:

$$f(t) = \frac{a_0}{2} + \sum_{n=1}^{\infty} a_n \sin(nt) + \sum_{n=1}^{\infty} b_n \cos(nt). \qquad (3.186)$$

So if we have a waveform we want to model we only need to find the coefficients $a_0, a_1, \ldots, a_n$ and $b_0, b_1, \ldots, b_n$ and the number of sines and cosines. We won't go into the derivation of these coefficients (or how to find the number of sines and cosines) here. The definition will be enough, as this is only meant to be a brief introduction to the Fourier series. For example, suppose we've found $a_1 = 0.5, a_4 = 2$ and $b_2 = 4$ and all the rest are 0; then the Fourier series is:

$$f(t) = 0.5\sin(\pi t) + 2\sin(4\pi t) + 4\cos(2\pi t)$$

which is represented by the following graph:

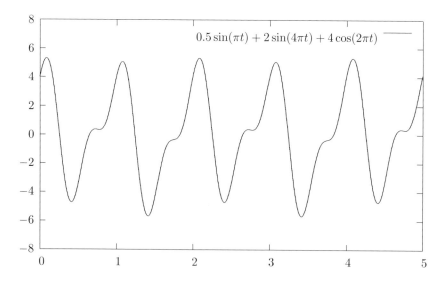

The function $f(t)$ is made up of the following waveforms, $0.5\sin(\pi t), 2\sin(4\pi t)$, and $4\cos(2\pi t)$. Again, it's helpful to look at them graphically:

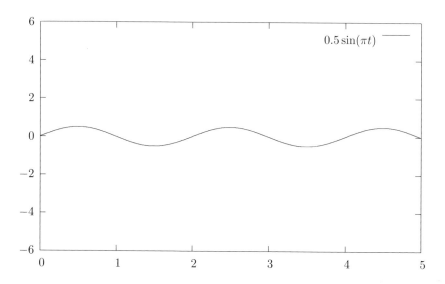

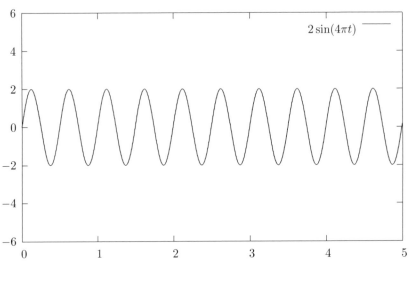

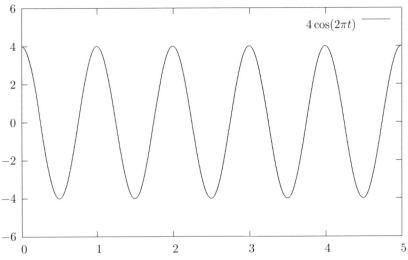

If we analyse the frequencies and amplitudes of the components of $f(t)$ we get the following results [34]:

Waveform	Sine Amplitude	Cosine Amplitude	Frequency
$0.5\sin(\pi t)$	$\frac{1}{2}$	0	2
$2\sin(4\pi t)$	2	0	$\frac{1}{2}$
$4\cos(2\pi t)$	0	4	1

We can also rewrite the sinusoids above as a sum of numbers in complex, exponential form.

3.9.2 The Discrete Fourier Transform

The DFT maps from a discrete, periodic sequence t_k to a set of coefficients representing the frequencies of the discrete sequence. The DFT takes an array of complex numbers as input and also outputs one. The number of elements in the array is governed by the *sampling rate* and the *length* of the waveform. Formally, the N complex numbers $t_0, ..., t_{N-1}$ are transformed into the N complex numbers $f_0, ..., f_N - 1$ according to the formula:

$$f_j = \sum_{k=0}^{N-1} t_k e^{-\frac{2\pi i}{N} jk} \qquad j = 0, \ldots, N-1. \qquad (3.187)$$

The DFT is a linear operator with an invertible matrix representation, so we can take the conversion back to its original form using:

$$t_k = \frac{1}{N} \sum_{j=0}^{N-1} f_j e^{\frac{2\pi i}{N} kj} \qquad k = 0, \ldots, N-1. \qquad (3.188)$$

We have chosen to represent our periodic functions as a sequence of sines and cosines. To use the above formulas they need to be converted to complex, exponential form. Because the sequence we are after is discrete we need to sample various points along the sequence. The sample rate N determines the accuracy of our transformation, with the lower bound on the sampling rate being found by applying *Nyquist's theorem* (which is beyond the scope of this tutorial).

Let's look at doing a DFT on $f(t) = 0.5\sin(\pi t) + 2\sin(4\pi t) + 4\cos(2\pi t)$ and adjust the sampling rate until we get an acceptable waveform in the frequency domain.

The graph below is just $f(t) = 0.5\sin(\pi t) + 2\sin(4\pi t) + 4\cos(2\pi t)$ with sampling points at the whole numbers $(1, 2, \ldots, N)$. As you can see if we only sample at this point we get no notion of a wave at all (see fig 3.10). Instead of adjusting the sample rate to be fractional, we just have to adjust the function slightly, which just makes the x-axis longer but retains our wave. The function now looks like this (see fig 3.11):

$$f(t) = 0.5\sin\left(\pi\frac{t}{2}\right) + 2\sin\left(4\pi\frac{t}{2}\right) + 4\cos\left(2\pi\frac{t}{2}\right).$$

So we are effectively sampling at twice the rate.

Below we show a sampling rate of 50 times the original rate and our waveform looks good, the function now looks like this (see fig 3.12):

$$f(t) = 0.5\sin\left(\pi\frac{t}{50}\right) + 2\sin\left(4\pi\frac{t}{50}\right) + 4\cos\left(2\pi\frac{t}{50}\right).$$

Finally, $f(t) = 0.5\sin(\pi\frac{t}{50}) + 2\sin(4\pi\frac{t}{50}) + 4\cos(2\pi\frac{t}{50})$ has been put through the DFT and now it is now in the frequency domain (see fig 3.13).

Later, in chapter 7, we'll see how the quantum analogue of the DFT (called the quantum fourier transform) can be used for quantum computing.

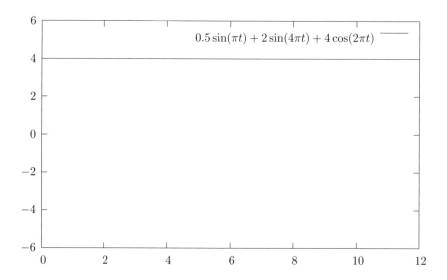

Fig. 3.10 $f(t) = 0.5\sin(\pi t) + 2\sin(4\pi t) + 4\cos(2\pi t)$.

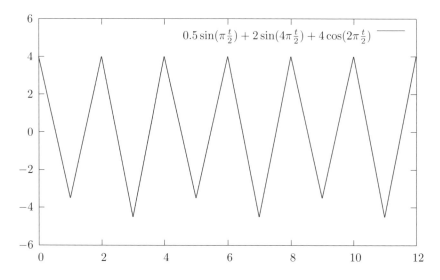

Fig. 3.11 $f(t) = 0.5\sin(\pi\frac{t}{2}) + 2\sin(4\pi\frac{t}{2}) + 4\cos(2\pi\frac{t}{2})$.

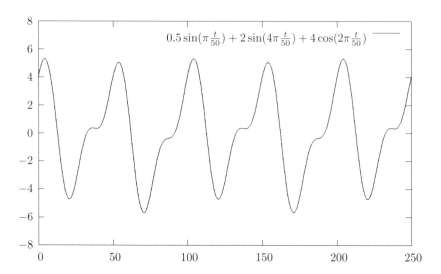

Fig. 3.12 $f(t) = 0.5\sin(\pi\frac{t}{50}) + 2\sin(4\pi\frac{t}{50}) + 4\cos(2\pi\frac{t}{50})$.

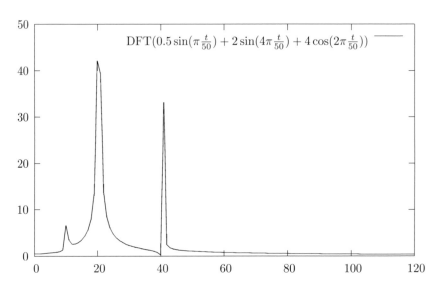

Fig. 3.13 $f(t) = 0.5\sin(\pi\frac{t}{50}) + 2\sin(4\pi\frac{t}{50}) + 4\cos(2\pi\frac{t}{50})$.

Chapter 4

Quantum Mechanics

Quantum mechanics is generally about the novel behaviour of very small things. At this scale matter becomes *quantised*, which means that it can be subdivided no more. Quantum mechanics has never been wrong: it explains why the stars shine, how matter is structured, the periodic table, and countless other phenomena. One day scientists hope to use quantum mechanics to explain everything, but at present the theory remains incomplete as it has not been successfully combined with classical theories of gravity.

Some strange effects happen at the quantum scale. The following effects are important for quantum computing:

- Superposition and interference
- Uncertainty
- Entanglement

This chapter is broken into two parts. In the first part we'll look briefly at the history of quantum mechanics. Then, in the second part we will examine some important concepts (like the ones above) of quantum mechanics and how they relate to quantum computing.

The main references used for this chapter are *Introducing Quantum Theory* by J.P. McEvoy and Oscar Zarate, and *Quantum Physics, Illusion or Reality* by Alastair Rae. Both of these are very accessible introductory books.

Fig. 4.1   James Clerk Maxwell and Isaac Newton.

4.1 History

4.1.1 *Classical Physics*

Classical physics roughly means pre-20th century physics, or pre-quantum physics. Two of the most important classical theories are *electromagnetism*, the unification of electricity and magnetism by James Clerk Maxwell, 1831–1879 (figure 4.1) and Isaac Newton's *mechanics*. Isaac Newton, 1642–1727 (figure 4.1) is arguably the most important scientist of all time due to the large body of work he produced that is still relevant today. Newton's contributions include work on gravity, optics and mathematics. Prior to this, Alhazen 965–1039, Nicolaus Copernicus, 1473–1543 and Galileo Galilei, 1564–1642 (figure 4.2) contributed greatly to the development of the modern scientific method (we might also include Leonardo Davinci, 1452–1519) by testing their theories with observation and experimentation.

Classical physics has a number of fundamental assumptions, they are:

Fig. 4.2   Nicolaus Copernicus and Galileo Galilei.

- The universe is a giant machine.
- *Cause and effect*, i.e. all nonuniform motion and action is caused by something (uniform motion doesn't need a cause, this is Galileo's principle of *inertia*).
- *Determinism*, if a state of motion is known now then because the universe is predictable, we can say exactly what it has been and what it will be at any time.
- Light is a wave that is completely described by a set of wave equations created by Maxwell. These are four equations that describe all electric and magnetic phenomena.
- Particles and waves exist, but they are distinct.
- We can measure any system to an arbitrary accuracy and correct for any errors caused by the measuring tool.

It's all proven, or is it? No, the above assumptions do not hold for quantum mechanics.

4.1.2   Important Concepts

In the lead up to quantum mechanics there are some important concepts from classical physics that we should look at. These are the concepts of atoms, thermodynamics, and statistical analysis.

Fig. 4.3   Democritus and John Dalton.

4.1.2.1   *Atoms*

Atoms are defined as indivisible parts of matter first postulated by Democritus, 460–370 BC (figure 4.3). The idea was dismissed as a waste of time by Aristotle, (384–322 BC) but two thousand years later the idea started gaining acceptance. The first major breakthrough was in 1806 when John

Dalton, 1766–1844 (figure 4.3) predicted properties of elements and compounds using the atomic concept.

4.1.2.2 *Thermodynamics*

Thermodynamics is the theory of heat energy. Heat is understood to be disordered energy; e.g. the heat energy in a gas is the kinetic energies of all the molecules. The temperature is a measure of how fast the molecules are traveling (if a gas or liquid; if solid, how fast they are vibrating about their fixed positions in the solid).

Thermodynamics is made up of four laws, the first two of which are important for us:

Fig. 4.4   Hermann von Helmholtz and Rudolf Clausius.

- **The first law of thermodynamics**

  In a closed system, whenever a certain amount of energy disappears in one place an equivalent amount must appear elsewhere in the system is some form.

  This law of conservation of energy was originally stated by Herman Von Helmholtz, 1824–1894 (figure 4.4).

- **The second law of thermodynamics**
  Rudolf Clausius, 1822–1888 (figure 4.4) called the previous law the first of two laws. He introduced a new concept, *entropy* which in terms of heat transfer is:

  The total entropy of a system increases when heat flows from a hot body to a cold one. Heat always flows from hot to cold [32].

So this implies that an isolated system's entropy is always increasing until the system reaches thermal equilibrium (i.e. all parts of the system are at the same temperature).

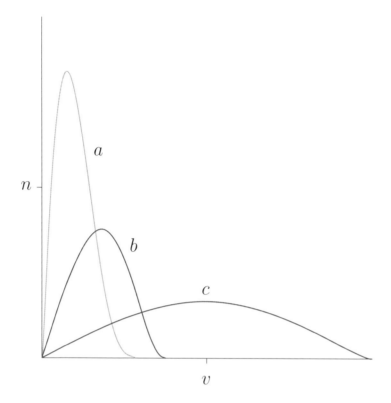

Fig. 4.5   Maxwell distribution.

4.1.3   *Statistical Mechanics*

In 1859, J.C. Maxwell, using the atomic model, came up with a way of statistically averaging the velocities of randomly chosen molecules of a gas in a closed system like a box (because it was impossible to track each one). The graph is shown in figure 4.5. Remember hotter molecules tend to go faster. The graph's n axis denotes the number of molecules (this particular graph shows CO_2) while the v axis denotes velocity. The letters a, b, and c represent molecules at 100K, 400K, and 1600K respectively.

In the 1870s Ludwig Boltzmann, 1844–1906 (figure 4.6) generalised the theory to any collection of entities that interact randomly, are independent, and are free to move. He rewrote the second law of thermodynamics to say:

As the energy in a system degrades the system's atoms become more disordered and there is an increase in entropy. to measure this disorder we consider the number of configurations or states that the collection of atoms can be in.

Fig. 4.6   Ludwig Boltzmann and Niels Bohr.

If this number is W then the entropy S is defined as:

$$S = k \log W \tag{4.1}$$

where k is Boltzmann's constant $k = 1.38 \times 10^{-23}\ J/K$. Here J stands for Joule, a measure of energy, and K stands for Kelvin, a unit of temperature. So the behaviour of "large things" could now be predicted by the average statistical behaviour of the their smaller parts, which is important for quantum mechanics. There also remains the probability that a *fluctuation* can occur, a statistical improbability that may seem nonsensical but nonetheless the theory must cater for it. For example, if we have a box containing a gas a fluctuation could be all particles of the gas randomly clumping together in one corner of the box.

4.1.4   *Important Experiments*

There are two major periods in the development of quantum theory. The first culminated in 1913 with the atomic model of Niels Bohr, 1885–1962 (figure 4.6) and ended in about 1924. This is called *old quantum theory*. The *new quantum theory* began in 1925. The old quantum theory was developed in some part to explain the results of three experiments which could not be explained by classical physics, they are:

- Black body radiation, and the ultraviolet catastrophe.
- The Photoelectric effect.

- Bright Line Spectra.

These experiments, and their subsequent explanations, are described in the next three sections.

4.1.4.1 *Black Body Radiation*

A *black body* absorbs all electromagnetic radiation (light) that falls on it and would appear black to an observer because it reflects no light. To determine the temperature of a black body we have to observe the radiation emitted from it.

Max Planck, 1858–1947, (figure 4.8) measured the distribution of radiation and energy over frequency in a *cavity*, a kind of oven with a little hole for a small amount of heat (light, radiation) to escape for observation. Because the radiation is confined in the cavity it settles down to an equilibrium distribution of the molecules in a gas. Planck found the frequency distributions to be similar to Maxwell's velocity distributions. The colour of the light emitted is dependent on the temperature, e.g. the element of your electric stove goes from red hot to white hot as the temperature increases. It didn't take long for physicists to apply a Maxwell-style statistical analysis to the waves of electromagnetic energy present in the cavity. The difference is that classical physics saw waves as continuous which means that more and more waves could be packed into a "box" as the wavelengths get smaller, i.e. the frequency gets higher. This means that as the temperature was raised the radiation should keep getting stronger and stronger indefinitely. This was called the ultraviolet catastrophe. If nature did indeed behave in this way you would get singed sitting in front of a fire by all the ultraviolet light coming out of it. Fortunately this doesn't occur so the catastrophe is not in nature but in classical physics which predicted something that doesn't happen.

The results of several experiments had given the correct frequency distributions and it was Max Planck who found a formula that matched the results. He couldn't find a classical solution, so grudgingly he used Boltzmann's version of the second law of thermodynamics. Planck imagined that the waves emitted from the black body were produced by a finite number of tiny oscillators (a kind of precursor to modern atoms). Eventually he had to divide the energy into finite chunks of a certain size to fit his own radiation formula, which finally gave us the first important formula for quantum mechanics:

$$E = hf \qquad (4.2)$$

where E is energy, f is frequency and h is *Planck's constant* which is:

$$h = 6.66260688 \times 10^{-34} J/s. \qquad (4.3)$$

where J/s = Joules per second.

Fig. 4.7   Albert Einstein and Johann Jakob Balmer.

4.1.5   *The Photoelectric Effect*

If a light is shone onto certain kinds of material (e.g. some metals or semi conductors) then electrons are released. When this effect was examined it was found that the results of the experiments did not agree with classical electromagnetic theory. Classical theory predicted that the energy of the released electron should depend on the intensity of the incident light wave. However it was found that the energy released was dependent not on intensity (an electron would come out no matter how low the intensity was) but on the frequency of the light.

Albert Einstein, 1879–1955 (figure 4.7) showed that if we look at the light as a collection of particles carrying energy proportional to the frequency (as given by Planck's law $E = hf$) and if those particles can transfer energy to electrons in a target metal then the experimental results could be explained. Put simply a light particle hits the metal's surface and its energy is transferred to an electron and becomes kinetic energy; so the electron is ejected from the metal. With different kinds of metals it can be easier or harder for electrons to escape.

4.1.6 *Bright Line Spectra*

When a solid is heated and emits white light then, if that light is concentrated and broken up into the separate colours by a prism, we get a rainbow-like spectrum (*continuous spectrum*) like the following:

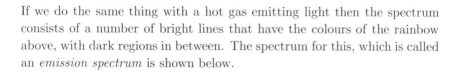

If we do the same thing with a hot gas emitting light then the spectrum consists of a number of bright lines that have the colours of the rainbow above, with dark regions in between. The spectrum for this, which is called an *emission spectrum* is shown below.

If a cold gas is placed between a hot solid emitting white light and the prism we get the inverse of the above. This is called an *absorbtion spectrum*, shown below.

The hot gas is emitting light at certain frequencies and example three shows us that the cold gas is absorbing light at the same frequencies. These lines are different for each element, and they allow us to determine the composition of a gas even at astronomical distances, by observing its spectrum.

In 1885 Johann Jakob Balmer, 1825–1898 (figure 4.7) derived a formula for the spectral lines of Hydrogen:

$$f = R \left(\frac{1}{n_f^2} - \frac{1}{n_i^2} \right) \tag{4.4}$$

where R is the *Rydberg constant* of 3.29163×10^{15} cycles/second and n_f and n_i are whole numbers. The trouble was that no one knew how to explain the formula. The explanation came in 1913 with Niels Bohr's atomic model.

4.1.7   *Proto Quantum Mechanics*

During the last part of the 19th century it was discovered that a number of "rays" were actually particles. One of these particles was the *electron*, discovered by Joseph J. Thomson, 1856–1940 (figure 4.8). In a study

Fig. 4.8   Max Planck and Joseph J. Thomson.

Fig. 4.9   Ernest Rutherford and Arnold Sommerfeld.

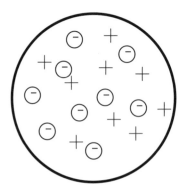

Fig. 4.10   Thomson's atomic model.

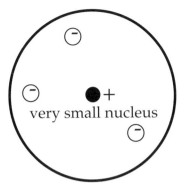

Fig. 4.11   Rutherford's atomic model.

of cathode ray tubes Thompson showed that electrically charged particles (electrons) are emitted when a wire is heated. Thomson went on to help develop the first model of the atom which had his (negatively charged) electrons contained within a positively charged sphere (figure 4.10).

This first atomic model was called the *Christmas pudding* model. Then, in 1907, Ernest Rutherford, 1871–1937 (figure 4.9), developed a new model, which was found by firing alpha particles (Helium ions) at gold foil and observing that, very occasionally, one would bounce back. This model had a tiny but massive nucleus surrounded by electrons (figure 4.11).

The new model was like a mini solar system with electrons orbiting the nucleus, but the atomic model was thought to still follow the rules of classical physics. However, according to classical electromagnetic theory an orbiting electron, subject to centripetal acceleration (the electron is attracted by the positively charged nucleus) would radiate energy and so rapidly spiral in towards the nucleus. But this did not happen: atoms were stable, and all the atoms of an element emitted the same line spectrum. To explain this Niels Bohr assumed that the atom could exist in only certain stationary states — stationary because, even if the electron was orbiting in such a state (and, later, this was questioned by Heisenberg) it would not radiate, despite what electromagnetic theory said. However, if the electron jumped from a stationary state to one of lower energy then the transmission was accompanied by the emission of a photon; vice versa there was absorption of light in going from a lower to a higher energy.

In this scheme there was a lowest stationary state, called the *ground state* below which the electron could not jump. This restored stability to the atom.

The frequency of the light emitted as a jump was given by Einstein's formula:

$$f = \frac{E}{h}$$

(4.5)

where E is the difference in the energies of the stationary states involved. These energies of the stationary states could be calculated from classical physics if one additional assumption was introduced: that the orbital angular momentum was an integer multiple of Planck's constant. Then the calculated frequencies were found to agree with those observed.

So Bohr developed a model based on stable orbital shells which only gave a certain number of shells to each atom. Bohr quantised electron orbits in units of Planck's constant. He gave us the first of several *quantum numbers* which are useful in quantum computing, the shell number, n (see figure 4.12).

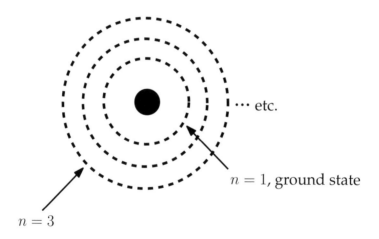

Fig. 4.12   Bohr's first atomic model.

Of particular interest are the ground state at $n = 1$ and the *excited state* at $n > 1$ of an atom. Bohr developed a formula for the radius of the electron orbits in a hydrogen atom:

$$r = \left(\frac{h^2}{4\pi^2 mq^2} \right) n^2$$

(4.6)

where r is the radius of the orbital, h is Planck's constant, and m and q

are the mass and charge of the electron. In real terms the value of r is 5.3 nanometers for $n = 1$.

Bohr went on with this model to derive the Balmer's formula for hydrogen by two postulates:

(1) Quantum angular momentum:

$$L = n \left(\frac{h}{2\pi} \right).$$ (4.7)

(2) A jump between orbitals will emit or absorb radiation by:

$$hf = E_i - E_f$$ (4.8)

where E_i is the initial energy of the electron and E_f is the final energy of the electron.

Although very close, it didn't quite match up to the spectral line data. Arnold Sommerfeld, 1868–1951 (figure 4.9) then proposed a new model with elliptical orbits and a new quantum number was added, k to deal with the shape of the orbit, Bohr then introduced quantum number m to explain the Zeeman effect which produced extra spectral lines when a magnetic field was applied to the atom (i.e. the direction the field was pointing).

It was soon discovered that m could not account for all the spectral lines produced by magnetic fields. Wolfgang Pauli, 1900–1958 (figure 4.13) hypothesised another quantum number to account for this. It was thought, but not accepted by Pauli that the electron was "spinning around" and it turns out that Pauli was right but the name stuck, so we still use *spin up* and *spin down* to describe this property of an electron. Pauli then described why electrons fill the higher energy levels and don't just occupy the ground state which we now call the *Pauli exclusion principle*. Niels Bohr went on to explain the periodic table in terms of orbital shells with the outermost shell being the valence shell that allows binding and the formation of molecules.

4.1.8   *The New Theory of Quantum Mechanics*

In 1909, a few years after demonstrating the photoelectric effect, Einstein used his photon hypothesis to obtain a simple derivation of Planck's black

Fig. 4.13   Wolfgang Pauli and Louis de Broglie.

body distribution. Planck himself had not gone as far as Einstein: he had indeed assumed that the transfer of energy between matter (the oscillators in the walls of the cavity) and radiation was quantised. But Planck had assumed the energy in the electromagnetic field, in the cavity, was continuously distributed, as in classical theory. By contrast, it was Einstein's hypothesis that the energy in the field itself was quantised: that for certain purposes, the field behaved like an ideal gas, not of molecules, but of photons, each with energy h times frequency, with the number of photons being proportional to the intensity. The clue to this was Einstein's observation that the high frequency part of Planck's distribution for black body radiation (described by *Wien's law*) could be derived by assuming a gas of photons and applying statistical mechanics to it. This was in contrast to the low frequency part (described by the *Rayleigh–Jeans law*) which could be successfully obtained using classical electromagnetic theory, i.e. assuming waves. So you had both particles and waves playing a part. Furthermore, Einstein looked at fluctuations of the energy about its average value, and observed that the formula obtained had two forms, one which you would get if light was made up of waves and the other if it was made up of particles. Hence we have wave-particle duality.

In 1924, Louis de Broglie, 1892–1987 (figure 4.13) extended the particle duality for light to all matter. He stated:

> The motion of a particle of any sort is associated with the propagation of a wave.

This is the idea of a pilot wave which guides a free particle's motion. de Broglie then suggested the idea of electron waves be extended to bound particles in atoms, meaning electrons move around the nucleus guided by pilot waves. So, again a duality, this time between de Broglie waves and

Fig. 4.14 Werner Heisenberg and Erwin Schrödinger.

Bohr's particles. de Broglie was able to show that Bohr's orbital radii could be obtained by fitting a whole number of waves around the nucleus. This gave an explanation of Bohr's angular momentum quantum condition (see section 4.1.6).

The new quantum theory was developed between June 1925 and June 1926. Werner Heisenberg, 1901–1976 (figure 4.14), using a totally different and simpler atomic model (one that did not use orbits) worked out a code to connect quantum numbers and spectra. He also discovered that quantum mechanics does not follow the *commutative law of multiplication* i.e $pq \neq qp$. When Max Born, 1882–1970 (figure 4.15) saw this he suggested that Heisenberg use matrices. This became matrix mechanics, and eventually all the spectral lines and quantum numbers were deduced for hydrogen. The first complete version of quantum mechanics was born. It's interesting to note that it was not observation or visualisation that was used to deduce the theory — but pure mathematics. Later we will see matrices cropping up in quantum computing.

At around the same time Erwin Schrödinger, 1887–1961 (figure 4.14) built on de Broglie's work on matter waves. He developed a wave equation (for which Ψ is the solution) for the core of bound electrons, as in the Hydrogen atom. It turns out that the results derived from this equation agree with the Bohr model. He then showed that Heisenberg's matrix mechanics and his wave mechanics were equivalent.

Max Born proposed that Ψ, the solution to Schrödinger's equation can be interpreted as a probability amplitude, not a real, physical value. The probability amplitude is a function of the electron's position (x, y, z) and, when squared, gives the probability of finding the electron in a unit volume at the point (x, y, z). This gives us a new, probabilistic atomic model,

Fig. 4.15   Max Born and Paul Dirac.

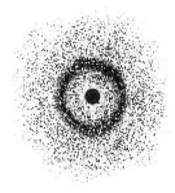

Fig. 4.16   Born's atomic model.

in which there is a high probability that the electron will be found in a particular orbital shell. A representation of the ground state of hydrogen is shown in figure 4.16 and at the places where the density of points is high there is a high probability of finding the particle. The linear nature of the wave equation means that if Ψ_1 and Ψ_2 are two solutions then so is $\Psi_1 + \Psi_2$, a superposition state (we'll look at superposition soon). This probabilistic interpretation of quantum mechanics implies the system is in both states until measured. Schrödinger was unhappy with the probabilistic interpretation (superposition) and created a scenario that would show it was false. This is called *Schrödinger's cat*, a paradox, which simply put refers to the situation of a cat being in both states of dead Ψ_1 and alive Ψ_2 until it is observed.

Paul Dirac, 1902–1984 (figure 4.15) developed a new approach to quantum mechanics and the bra-ket notation we use for quantum computing (see section 4.2.3).

In 1927 Heisenberg made his second major discovery, the *Heisenberg uncertainty principle* which relates to the position and momentum of a particle. It states that the more accurate our knowledge of a particle's position, the more inaccurate our knowledge of its momentum will be and vice versa. The uncertainty is due to the uncontrollable effect on the particle of any attempt to observe it (because of the quantum interaction; see 4.25) and perhaps to a genuine lack of the particle having precise properties. This signalled the breakdown of determinism.

Now back to Niels Bohr. In 1927 Niels Bohr described the concept of complementarity: it depends on what type of measurement operations you are using to look at the system as to whether it behaves like a particle or a wave. He then put together various aspects of the work by Heisenberg, Schrödinger and Born and concluded that the properties of a system (such as position and momentum) are *undefined* having only potential values with certain probabilities of being measured. This became know as the *Copenhagen interpretation* of quantum mechanics.

Einstein did not like the Copenhagen interpretation and, for a good deal of time, Einstein kept trying to refute it by thought experiment, but Bohr always had an answer. But in 1935 Einstein raised an issue that was to later have profound implications for quantum computation and lead to the phenomenon we now call entanglement, a concept we'll look at a little later.

4.2 Important Principles for Quantum Computing

The main parts of quantum mechanics that are important for quantum computing are:

- Linear algebra.
- Superposition.
- Dirac notation.
- Representing information.
- Uncertainty.
- Entanglement.
- The 4 postulates of quantum mechanics.

4.2.1 *Linear Algebra*

Quantum mechanics leans heavily on linear algebra. Some of the concepts of quantum mechanics come from the mathematical formalism, not thought experiments — that's what can give rise to counter-intuitive conclusions.

4.2.2 *Superposition*

Superposition means a system can be in two or more of its states simultaneously. For example a single particle can be traveling along two different paths at once. This implies that the particle has wave-like properties, which can mean that the waves from the different paths can *interfere* with each other. Interference can cause the particle to act in ways that are impossible to explain without these wave-like properties.

The ability for the particle to be in a superposition is where we get the parallel nature of quantum computing: if each of the states corresponds to a different value then, if we have a superposition of such states and act on the system, we effectively act on all the states simultaneously.

4.2.2.1 *An Example With Silvered Mirrors*

Superposition can be explained by way of a simple example using silvered and half silvered mirrors [3].

A half silvered mirror reflects half of the light that hits it and transmits the other half of the light through it (figure 4.17). If we send a single photon through this system then this gives us a 50% chance of the light hitting detector 1 and a 50% chance of hitting detector 2. It is tempting to think that the light takes one or the other path, but in fact it takes both! It's just that the photo detector that *measures* the photon first breaks the superposition, so it's the detectors that cause the randomness, not the half silvered mirror. This can be demonstrated by adding in some fully silvered mirrors and bouncing both parts of the superposed photon (which is at this point is in two places at once) so that they meet and interfere with each other at their meeting point. If another half silvered mirror (figure 4.18) is placed at this meeting point and if light was just particle-like we would expect that the light would behave as before (going either way with 50% probability), but the interference (like wave interference when two stones are thrown into a pond near each other simultaneously) causes the photon to always be detected by detector 1. A third example (figure 4.19) shows clearly that the photons travel both paths because blocking one path will break the superposition and stop the interference.

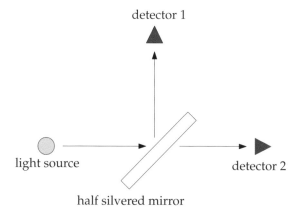

Fig. 4.17 Uncertainty.

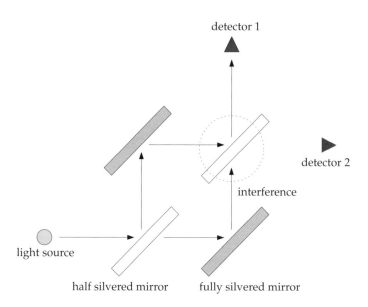

Fig. 4.18 Superposition 1.

4.2.3 *Dirac Notation*

As described in the previous chapter Dirac notation is used for quantum computing. We can represent the states of a quantum system as kets. For example, an electron's spin can be represented as $|0\rangle$ = spin up and $|1\rangle$ as spin down.

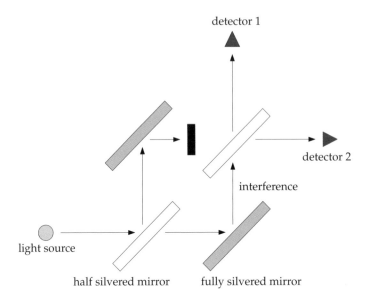

Fig. 4.19   Superposition 2.

The electron can be thought of as a little magnet, the effect of a charged particle spinning on its axis. When we pass a horizontally traveling electron through an inhomogeneous magnetic field, in, say, the vertical direction, the electron either goes up or down. If we then repeat this with the up electron it goes up, while with the down electron it goes down. We say the up electron after the first measurement is in the state $|0\rangle$ and the down electron is in state $|1\rangle$. But, if we take the up electron and pass it through a horizontal field it comes out on one side 50% of the time and on the other side 50% of the time. If we represent these two states as $|+\rangle$ and $|-\rangle$ we can say that the up spin electron was in a superposition of the two states $|+\rangle$ and $|-\rangle$ such that, when we make a measurement with the field horizontal we project the electron into one or the other of the two states, with equal probabilities of $\frac{1}{2}$.

4.2.4   *Representing Information*

Quantum mechanical information can be physically realised in many ways. To have something analogous to a classical bit we need a quantum mechanical system with two states only, when measured. We have just seen two examples: electron spin and photon direction. Two more methods for representing binary information in a way that is capable of exhibiting quantum effects (e.g. entanglement and superposition) are: *polarisation of photons* and *nuclear spins*.

We examine various physical implementations of these *quantum bits* (qubits) in chapter 8.

4.2.5   *Uncertainty*

The quantum world is irreducibly small so it's impossible to measure a quantum system without having an effect on that system as our measurement device is also quantum mechanical. As a result there is no way of accurately predicting all of the properties of a particle. There is a trade off — the properties occur in complementary pairs (like position and momentum, or vertical spin and horizontal spin) and if we know one property with a high degree of certainty then we must know almost nothing about the other property. That unknown property's behaviour is essentially random. An example of this is a particle's position and velocity: if we know exactly where it is then we know nothing about how fast it is going. This indeterminacy is exploited in quantum cryptography (see chapter 7).

It has been postulated (and currently accepted) that particles in fact DO NOT have defined values for unknown properties until they are measured. This is like saying that something does not exist until it is looked at.

4.2.6   *Entanglement*

In 1935 Einstein (along with colleagues Podolski and Rosen) demonstrated a paradox (named *EPR* after them) in an attempt to refute the undefined nature of quantum systems. The results of their experiment seemed to show that quantum systems were defined, having *local state* BEFORE measurement. Although the original hypothesis was later proven wrong (i.e. it was proven that quantum systems do not have local state before measurement), the effect they demonstrated was still important, and later became known as *entanglement*.

Entanglement is the ability for pairs of particles to *interact* over any distance instantaneously. Particles don't exactly communicate, but there is a statistical *correlation* between results of measurements on each particle that is hard to understand using classical physics. To become entangled, two particles are allowed to interact; they then separate and, on measuring say, the velocity of one of them (regardless of the distance between them), we can be sure of the value of velocity of the other one (before it is measured). The reason we say that they communicate instantaneously is because they store no local state [32] and only have well defined state once they are measured. Because of this limitation particles can't be used to transmit classical messages faster than the speed of light as we only know the states upon measurement. Entanglement has applications in a wide variety of quantum algorithms and machinery, some of which we'll look at later.

As stated before, it has been proven that entangled particles have no local state; this is explained in section 6.7.

Chapter 5

Quantum Computing

5.1 Elements of Quantum Computing

5.1.1 *Introduction*

Generally we'll think of a quantum computer as a classical computer with a quantum circuit attached to it with some kind of interface between conventional and quantum logic. Since there are only a few things a quantum computer does better than a classical computer it makes sense to do the bulk of the processing on the classical machine.

This section borrows heavily from [31] and [14] so the use of individual citations for these references has been dropped.

5.1.2 *History*

In 1982 Richard Feynman theorised that classic computation could be dramatically improved by quantum effects. Building on this, David Deutsch developed the basis for quantum computing between 1984 and 1985. The next major breakthrough came in 1994 when Peter Shor described a method to factor large numbers in quantum poly-time (which breaks RSA encryption). This became known as Shor's algorithm. At around the same time the quantum complexity classes were developed and the quantum Turing machine was described. Then in 1996 Lov Grover developed a fast database search algorithm (known as Grover's algorithm). The first prototypes of quantum computers were also built in 1996. In 1997 quantum error correction techniques were developed at Bell labs and IBM. Physical implementations of quantum computers improved with a three qubit machine in 1999 and a seven qubit machine in 2000.

5.1.3 *Bits and Qubits*

This section is about the nuts and bolts of quantum computing. It describes qubits, gates, and circuits.

Quantum computers perform operations on qubits which are analogous to conventional bits (see below) but they have an additional property in that they can be in a superposition. A quantum register with 3 qubits can store 8 numbers in superposition simultaneously [3] and a 250 qubit register holds more numbers (superposed) than there are atoms in the universe! [17].

The amount of information stored during the computational phase is essentially infinite - its just that we can't get at it. The inaccessibility of the information is related to quantum measurement: when we attempt to readout a superposition state holding many values the state collapses and we get only one value (the rest get lost). This is tantalising but, in some cases, can be made to work to our computational advantage.

5.1.3.1 *Single Qubits*

Classical computers use two discrete states (e.g. states of charging of a capacitor) to represent a unit of information. This state is called a binary digit (or bit for short). A bit has the following two values:

$$0 \text{ and } 1.$$

There is no intermediate state between them, i.e. the value of the bit cannot be in a superposition.

Quantum bits, or *qubits*, can on the other hand be in a state "between" 0 and 1, but only during the computational phase of a quantum operation. When measured, a qubit can become either:

$$|0\rangle \text{ or } |1\rangle$$

i.e. we readout 0 or 1. This is the same as saying a spin particle can be in a superposition state but, when measured, it shows only one value (see chapter 4).

The $|\rangle$ symbolic notation is part of the Dirac notation (see chapters 3 and 4). In terms of the above it essentially means the same thing as 0 and 1 (this is explained a little further on), just like a classical bit. Generally, a qubit's state during the computational phase is represented by a linear combination of states otherwise called a superposition state.

$$\alpha|0\rangle + \beta|1\rangle.$$

Here α and β are the probability amplitudes. They can be used to calculate the probabilities of the system jumping into $|0\rangle$ or $|1\rangle$ following a measurement or readout operation. There may be, say a 25% chance a 0 is measured and a 75% chance a 1 is measured. The percentages must add to 100%. In terms of their representation qubits must satisfy:

$$|\alpha|^2 + |\beta|^2 = 1. \tag{5.1}$$

This is the same thing as saying the probabilities add to 100%.

Once the qubit is measured it will remain in that state if the same measurement is repeated provided the system remains closed between measurements (see chapter 4). The probability that the qubit's state, when in a superposition, will collapse to states $|0\rangle$ or $|1\rangle$ is:

$$|\alpha|^2 \text{ for } |0\rangle$$

and

$$|\beta|^2 \text{ for } |1\rangle.$$

$|0\rangle$ and $|1\rangle$ are actually vectors, and are called the computational basis states that form an orthonormal basis for the vector space $\mathbb{C}^2$.

The state vector $|\Psi\rangle$ of a quantum system describes the state at any point in time of the entire system. Our state vector in the case of one qubit is:

$$|\Psi\rangle = \alpha|0\rangle + \beta|1\rangle. \tag{5.2}$$

The α and β might vary with time as the state evolves during the computation but the sum of the squares of α and β must always must be equal to 1.

Quantum computing also commonly uses $\frac{1}{\sqrt{2}}(|0\rangle+|1\rangle)$ and $\frac{1}{\sqrt{2}}(|0\rangle-|1\rangle)$ as a basis for $\mathbb{C}^2$, which is often shortened to just $|+\rangle$, and $|-\rangle$. These bases are sometimes represented with arrows which are described below, and are referred to as *rectilinear* and *diagonal* which can refer to say the polarisation of a photon. You may find these notational conventions being used:

$$|0\rangle = |\rightarrow\rangle. \tag{5.3}$$

$$|1\rangle = |\uparrow\rangle. \tag{5.4}$$

$$\frac{1}{\sqrt{2}}(|0\rangle + |1\rangle) = |+\rangle = |\nearrow\rangle. \tag{5.5}$$

$$\frac{1}{\sqrt{2}}(|0\rangle - |1\rangle) = |-\rangle = |\nwarrow\rangle. \tag{5.6}$$

Some examples of measurement probabilities are on the next page.

Example Measurement probabilities.

$$|\Psi\rangle = \frac{1}{\sqrt{2}}|0\rangle + \frac{1}{\sqrt{2}}|1\rangle.$$

The probability of measuring a $|0\rangle$ is:

$$\left(\left|\frac{1}{\sqrt{2}}\right|\right)^2 = \frac{1}{2}.$$

The probability of measuring a $|1\rangle$ is:

$$\left(\left|\frac{1}{\sqrt{2}}\right|\right)^2 = \frac{1}{2}.$$

So 50% of the time we'll measure a $|0\rangle$ and 50% of the time we'll measure a $|1\rangle$.

Example More measurement probabilities.

$$|\Psi\rangle = \frac{\sqrt{3}}{2}|0\rangle - \frac{1}{2}|1\rangle.$$

The probability of measuring a $|0\rangle$ is:

$$\left(\left|\frac{\sqrt{3}}{2}\right|\right)^2 = \frac{3}{4}.$$

The probability of measuring a $|1\rangle$ is:

$$\left(\left|\frac{1}{2}\right|\right)^2 = \frac{1}{4}.$$

So 75% of the time we'll measure a $|0\rangle$ and 25% of the time we'll measure a $|1\rangle$.

The sign in the middle of the two values can change, which affects the internal evolution of the qubit, not the outcome of a measurement. When measuring in the basis $\{|0\rangle, |1\rangle\}$ the sign is actually the *relative phase* of the qubit. So,

$$\alpha|0\rangle + \beta|1\rangle$$

and

$$\alpha|0\rangle - \beta|1\rangle$$

have the same output values and probabilities but behave differently during the computational phase. Formally we say they differ by a relative phase factor. So in the case of the qubits above they differ by a phase factor of -1. It is important to note that while,

$$\frac{|0\rangle + |1\rangle}{\sqrt{2}} = \frac{|1\rangle + |0\rangle}{\sqrt{2}}$$

which is equivalent to,

$$|0\rangle = \begin{bmatrix} \frac{|1\rangle}{\sqrt{2}} \\ \frac{|1\rangle}{\sqrt{2}} \end{bmatrix} \tag{5.7}$$

the same does not apply for a minus sign,

$$\frac{|0\rangle - |1\rangle}{\sqrt{2}} \neq \frac{|1\rangle - |0\rangle}{\sqrt{2}}$$

where the vector representing the left state is,

$$|0\rangle = \begin{bmatrix} \frac{|1\rangle}{\sqrt{2}} \\ \frac{-|1\rangle}{\sqrt{2}} \end{bmatrix} \tag{5.8}$$

which differs from the vector representing the right state,

$$|0\rangle = \begin{bmatrix} \frac{-|1\rangle}{\sqrt{2}} \\ \frac{|1\rangle}{\sqrt{2}} \end{bmatrix}. \tag{5.9}$$

The other type of phase is called *global phase*. Two states can differ by a global phase factor and still be considered the same, as the global phase factor is not observable. One reason for this is that the probabilities for the outcomes $|\alpha|$ and $|\beta|$ are unaffected if α and β are each multiplied by the same complex number of magnitude 1. Likewise the relative phase (which figures in interference effects) is unaffected if α and β are multiplied by a common phase factor. What this means is that if we have a state on n qubits we can put a complex factor in front of the entire state to make it more readable. This is best described by an example (below).

Example Global phase.

$$|\Psi\rangle = \frac{-i}{\sqrt{2}}|0\rangle + \frac{1}{\sqrt{2}}|1\rangle.$$

can be rewritten as:

$$|\Psi\rangle = -i\left(\frac{1}{\sqrt{2}}|0\rangle - \frac{i}{\sqrt{2}}|1\rangle\right).$$

Remembering that $-i \times -i = -1$ we say the factor at the front of our state vector $(-i)$ is a global phase factor. We can also say here that because $-i = e^{-i\frac{\pi}{2}}$ we have a phase of $-\frac{\pi}{2}$.

Example More global phase.

$$|\Psi\rangle = \frac{1}{2}(-|00\rangle + |01\rangle - |10\rangle + |11\rangle).$$

can be rewritten as:

$$|\Psi\rangle = (-1)\frac{1}{2}(|00\rangle - |01\rangle + |10\rangle - |11\rangle).$$

5.1.3.2   *The Ket $|\rangle$*

Part of the Dirac notation is the ket ($|\rangle$). The ket is just a notation for a vector. The state of a single qubit is a unit vector in $\mathbb{C}^2$. So,

$$\begin{bmatrix} \alpha \\ \beta \end{bmatrix}$$

is a vector, and is written as:

$$\alpha|0\rangle + \beta|1\rangle$$

with

$$|0\rangle = \begin{bmatrix} 1 \\ 0 \end{bmatrix} \tag{5.10}$$

and

$$|1\rangle = \begin{bmatrix} 0 \\ 1 \end{bmatrix}. \tag{5.11}$$

5.1.3.3   *Two Dimensional Qubit Visualisation*

Single qubits can be represented in value and relative phase in two dimensions by the diagram in figure 5.1 which is similar to the way we represent polar coordinates for complex numbers. The graph shows the general form of 2D qubit representation where $a = \frac{1}{\sqrt{2}}|0\rangle + \frac{1}{\sqrt{2}}|1\rangle$ and $b = \frac{1}{\sqrt{2}}|0\rangle - \frac{1}{\sqrt{2}}|1\rangle$.

This diagram is ok for real numbered values of α and β but cannot accurately depict all the possible states of a qubit. For this we need three dimensions.

5.1.3.4   *Three Dimensional Qubit Visualisation — The Bloch Sphere*

The *Bloch sphere* is a tool with which the state of single qubit can be viewed in three dimensions and is useful for visualising all single qubit operations.

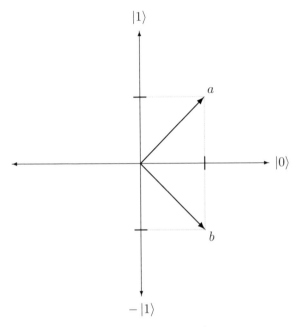

Fig. 5.1   2D qubit representations.

We can say that the state of a single qubit can be written as:

$$|\Psi\rangle = e^{i\gamma}(\cos\frac{\theta}{2}|0\rangle + e^{i\varphi}\sin\frac{\theta}{2}|1\rangle). \tag{5.12}$$

We can ignore the global phase factor in front so $|\Psi\rangle$ becomes:

$$|\Psi\rangle = \cos\frac{\theta}{2}|0\rangle + e^{i\varphi}\sin\frac{\theta}{2}|1\rangle. \tag{5.13}$$

So, in terms of the angle θ and φ the Bloch sphere looks like this:

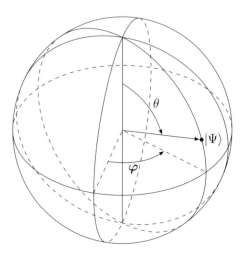

What's probably more helpful at this stage is to see where all of the potential states of a qubit lie on the Bloch sphere. This is shown below with the points $\hat{x}, \hat{y}$, and $\hat{z}$ labeling each positive axis:

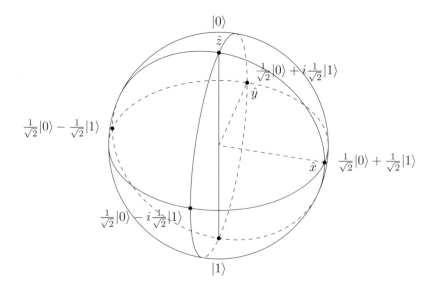

As stated in chapter 4, individual qubits can be physically realised using various quantum two state systems, here are a few ways this can be done:

- Polarisations of a photon.
- Nuclear spins.
- Ground and excited states of an atom (i.e. the energy level, or orbital).

We now look at the equivalent of a register: i.e. a composite system of qubits, e.g. Ions in a trap (see chapter 8).

5.1.3.5 *Multiple Qubits*

The potential amount of information available during the computational phase grows exponentially with the size of the system, i.e. the number of qubits (an example is a Qubyte, which consists of eight qubits). This is because if we have n qubits the number of basis states is 2^n. E.g. if we have two qubits, forming a quantum register then there are four ($= 2^2$) computational basis states: forming,

$$|00\rangle, |01\rangle, |10\rangle, \text{ and } |11\rangle. \tag{5.14}$$

Here $|01\rangle$ means that qubit 1 is in state $|0\rangle$ and qubit 2 is in state $|1\rangle$, etc. We actually have $|01\rangle = |0\rangle \otimes |1\rangle$, where $\otimes$ is the tensor product (see below).

Like a single qubit, the two qubit registers can exist in a superposition of the four states (below we change the notation for the complex coefficients, i.e. probability amplitudes):

$$|\Psi\rangle = \alpha_0|00\rangle + \alpha_1|01\rangle + \alpha_2|10\rangle + \alpha_3|11\rangle. \tag{5.15}$$

Again all of the probabilities must sum to 1. Formally for the general case of n qubits this is can be written as:

$$\sum_{i=0}^{2^n-1} |\alpha_i|^2 = 1. \tag{5.16}$$

Example $n = 5$ (5 qubits). We can have up to 32 $(= 2^5)$ basis states in a superposition.

$$\Psi = \alpha_0|00000\rangle + \alpha_1|00001\rangle + ... + \alpha_{2^n-1}|11111\rangle.$$

We do not have to represent values with 0s and 1s. A *qudit* has the following format in $\mathbb{C}^N$:

$$\Psi = \alpha_0|0\rangle + \alpha_1|1\rangle + \alpha_2|2\rangle + ... + \alpha_{n-1}|N-1\rangle = \begin{bmatrix} \alpha_0 \\ \alpha_1 \\ \alpha_2 \\ \vdots \\ \alpha_N - 1 \end{bmatrix}. \qquad (5.17)$$

If $N = 2^n$ we require an n qubit register.

5.1.3.6   *Tensor Products*

A decomposition into single qubits of a multi-qubit system can be represented by a tensor product, $\otimes$.

Example   Decomposition using a tensor product.

$$\frac{1}{2}(|00\rangle + |01\rangle + |10\rangle + |11\rangle) = \frac{1}{\sqrt{2}}(|0\rangle + |1\rangle) \otimes \frac{1}{\sqrt{2}}(|0\rangle + |1\rangle).$$

A tensor product can also be used to combine different qubits.

$$(\alpha_0|0\rangle + \alpha_1|1\rangle) \otimes (\beta_0|0\rangle + \beta_1|1\rangle) = \alpha_0\beta_0|00\rangle + \alpha_0\beta_1|01\rangle + \alpha_1\beta_0|10\rangle + \alpha_1\beta_1|11\rangle. \qquad (5.18)$$

5.1.3.7   *Partial Measurement*

We can measure a subset of an n-qubit system. i.e. we do not have to get readouts on all the qubits (some can be left unmeasured). We'll first consider nonentangled states. The simplest way to measure a subset of states is shown in the following example with two qubits.

Example Measuring the first bit in a two qubit system.

1. We prepare a quantum system in the following state. The qubit we are going to measure is bolded. A nonentangled state would mean that all the probability amplitudes are nonzero:

$$\Psi = \alpha_0|\mathbf{0}0\rangle + \alpha_1|\mathbf{0}1\rangle + \alpha_2|\mathbf{1}0\rangle + \alpha_3|\mathbf{1}1\rangle.$$

2. We now measure, so the probability of it being 0 is:

$$p_0 = |\alpha_0|^2 + |\alpha_1|^2$$

and, the probability of it being 1 is:

$$p_1 = |\alpha_2|^2 + |\alpha_3|^2.$$

3. If we measured a $|0\rangle$ the post measurement state is:

$$|0\rangle \otimes \frac{\alpha_0|0\rangle + \alpha_1|1\rangle}{\sqrt{|\alpha_0|^2 + |\alpha_1|^2}}$$

(i.e. we project onto the $\{|00\rangle, |01\rangle\}$ subspace and the α_2 and α_3 terms drop out). Similarly, if we measured a $|1\rangle$ measured the post measurement state is:

$$|1\rangle \otimes \frac{\alpha_2|0\rangle + \alpha_3|1\rangle}{\sqrt{|\alpha_2|^2 + |\alpha_3|^2}}.$$

We can do the same for qubit two, the probability of qubit two being a $|0\rangle$ is:

$$p_0 = |\alpha_0|^2 + |\alpha_2|^2$$

and its post measurement state would be:

$$\frac{\alpha_0|0\rangle + \alpha_2|1\rangle}{\sqrt{|\alpha_0|^2 + |\alpha_2|^2}} \otimes |0\rangle.$$

This logic can be extended to n qubits.

Quantum measurement can be described as a set $\{M_m\}$ of linear operators with $1 \leq m \leq n$ where n is the number of possible outcomes. For a single qubit with an orthonormal basis of $|0\rangle$ and $|1\rangle$ we can define measurement operators $M_0 = |0\rangle\langle 0|$ and $M_1 = |1\rangle\langle 1|$ (which are also both projectors).

If we have a system in state $|\Psi\rangle$ then outcome m has a probability of:

$$p_m = \langle \Psi | M_m^\dagger M_m | \Psi \rangle. \tag{5.19}$$

If the outcome is m then the state collapses to:

$$\frac{M_m|\Psi\rangle}{\sqrt{\langle \Psi | M_m^\dagger M_m | \Psi \rangle}} \,. \tag{5.20}$$

Example Another way of looking at measuring the first bit in a two qubit system.

$$|\Psi\rangle = \frac{1}{\sqrt{30}}|00\rangle + \frac{2}{\sqrt{30}}|01\rangle + \frac{3}{\sqrt{30}}|10\rangle + \frac{4}{\sqrt{30}}|11\rangle$$

When we measure qubit one the resulting states would look like this (unnormalised) for measuring a $|0\rangle$:

$$|\Psi\rangle = |0\rangle \otimes \left(\frac{1}{\sqrt{30}}|0\rangle + \frac{2}{\sqrt{30}}|1\rangle\right)$$

and for measuring a $|1\rangle$:

$$|\Psi\rangle = |1\rangle \otimes \left(\frac{3}{\sqrt{30}}|0\rangle + \frac{4}{\sqrt{30}}|1\rangle\right).$$

Now we must make sure that the second qubit is normalised, so we multiply it by a factor:

$$|\Psi\rangle = \frac{\sqrt{5}}{\sqrt{30}}|0\rangle \otimes \left(\frac{1}{\sqrt{5}}|0\rangle + \frac{2}{\sqrt{5}}|1\rangle\right) + \frac{5}{\sqrt{30}}|1\rangle \otimes \left(\frac{3}{5}|0\rangle + \frac{4}{5}|1\rangle.\right)$$

This gives us $\left|\frac{\sqrt{5}}{\sqrt{30}}\right|^2 = \frac{1}{6}$ probability of measuring a $|0\rangle$ and a $\left|\frac{5}{\sqrt{30}}\right|^2 = \frac{5}{6}$ probability of measuring a $|1\rangle$. So if we measure a $|0\rangle$ then our post measurement state is:

$$|\Psi\rangle = |0\rangle \otimes \left(\frac{1}{\sqrt{5}}|0\rangle + \frac{2}{\sqrt{5}}|1\rangle\right)$$

and if we measure a $|1\rangle$ then our post measurement state is:

$$|\Psi\rangle = |1\rangle \otimes \left(\frac{3}{5}|0\rangle + \frac{4}{5}|1\rangle\right).$$

Example Measurement of qubit one in a two qubit system using a simple projector.

$$|\Psi\rangle = \frac{1}{\sqrt{30}}|00\rangle + \frac{2}{\sqrt{30}}|01\rangle + \frac{3}{\sqrt{30}}|10\rangle + \frac{4}{\sqrt{30}}|11\rangle.$$

We find the probability of measuring a $|0\rangle$ by using a projector $|00\rangle\langle00| + |01\rangle\langle01|$:

$$(|00\rangle\langle00| + |01\rangle\langle01|)\left(\frac{1}{\sqrt{30}}|00\rangle + \frac{2}{\sqrt{30}}|01\rangle + \frac{3}{\sqrt{30}}|10\rangle + \frac{4}{\sqrt{30}}|11\rangle\right)$$

$$= (|00\rangle\langle00| + |01\rangle\langle01|)\,|\Psi\rangle$$

$$= \frac{1}{\sqrt{30}}|00\rangle + \frac{2}{\sqrt{30}}|01\rangle.$$

Say we change our *measurement basis* to $\{|0\rangle, |1\rangle\}$. Then we can represent projectors P_0 and P_1 as $|0\rangle\langle0|$ and $|1\rangle\langle1|$ respectively. We measure the probability of the first qubit being 0 by using P_0 on qubit one and I on qubit two, i.e. $P_0 \otimes I$. If we wanted to measure the probability of the first qubit being 1 then we would use $P_1 \otimes I$.

$$p_0 = \langle\Psi|P_0 \otimes I|\Psi\rangle$$

$$= \langle\Psi|\,|0\rangle\langle0| \otimes I|\Psi\rangle$$

$$= \langle\Psi|\frac{1}{\sqrt{30}}|00\rangle + \frac{2}{\sqrt{30}}|01\rangle$$

$$= \frac{1}{6}$$

and this gives us a post-measurement state of:

$$|\Psi'\rangle = \frac{P_0 \otimes I|\Psi\rangle}{\sqrt{\langle\Psi|P_0 \otimes I|\Psi\rangle}}$$

$$= \frac{\frac{1}{\sqrt{30}}|00\rangle + \frac{2}{\sqrt{30}}|01\rangle}{\sqrt{\frac{1}{6}}}$$

$$= |0\rangle \otimes \left(\frac{\frac{1}{\sqrt{30}}|0\rangle + \frac{2}{\sqrt{30}}|1\rangle}{\sqrt{\frac{1}{6}}}\right)$$

$$= |0\rangle \otimes \left(\frac{1}{\sqrt{5}}|0\rangle + \frac{2}{\sqrt{5}}|1\rangle\right).$$

Properties:

All probabilities sum to 1,

$$\sum_{i=1}^{m} p_i = \sum_{i=1}^{m} \langle \Psi | M_i^\dagger M_i | \Psi \rangle = 1 .$$

This is the result of the completeness equation, i.e.

$$\sum_{i=1}^{m} M_i^\dagger M_i = I . \tag{5.21}$$

Note: our basis needs to be orthogonal, otherwise we can't reliably distinguish between two basis states $|u\rangle$ and $|v\rangle$, i.e. $\langle u|v\rangle \neq 0$ means $|u\rangle$ and $|v\rangle$ are not orthogonal.

5.1.3.8 *Projective Measurements*

Projective measurements are a means by which we can accomplish two tasks, they are:

(1) Apply a unitary transform to $|\Psi\rangle$.
(2) Measure $|\Psi\rangle$.

So we need a unitary transform U and $|\Psi\rangle$ to perform a measurement. The unitary transform is called the *observable*, which is denoted here by O_M.

First we need to find the spectral decomposition of O_M (Z for example). For O_M we have:

$$O_M = \sum_{m} m P_m \tag{5.22}$$

where m is each eigenvalue and P_m is a projector made up of $P_m = |m\rangle\langle m|$.

5.1.4 **Entangled States**

Subatomic particles can be entangled. This means that they are connected, regardless of distance. Their effect on each other upon measurement is instantaneous. This can be useful for computational purposes.

Consider the following state (which is not entangled):

$$\frac{1}{\sqrt{2}} (|00\rangle + |01\rangle)$$

it can be expanded to:

$$\frac{1}{\sqrt{2}}|00\rangle + \frac{1}{\sqrt{2}}|01\rangle + 0|10\rangle + 0|11\rangle.$$

Upon measuring the first qubit (a partial measurement) we get 0 100% of the time and the state of the second qubit becomes:

$$\frac{1}{\sqrt{2}}|0\rangle + \frac{1}{\sqrt{2}}|1\rangle$$

giving us equal probability for a 0 or a 1.

If we try this on an entangled state (in this case an EPR pair or Bell state, detailed in section 6.7) we find that the results for the qubits are correlated.

Example   Consider:

$$\frac{1}{\sqrt{2}}(|00\rangle + |11\rangle) .$$

When expanded this is:

$$\frac{1}{\sqrt{2}}|00\rangle + 0|01\rangle + 0|10\rangle + \frac{1}{\sqrt{2}}|11\rangle.$$

Measuring the first qubit gives us $|00\rangle$ 50% of the time and $|11\rangle$ 50% of the time. So the second qubit is always the same as the first, i.e. we get two qubit values for the price of one measurement.

This type of correlation can be used in a variety of ways in application to the first or second qubit to give us correlations that are strongly statistically connected. This is a distinct advantage over classical computation. Measuring entangled states accounts for the correlations between them.

A number of qubits entangled together can be used as a quantum register (e.g. a qubyte). Quantum gates generally act on individual qubits within a register. Some mechanism will change the physical property of the qubit(s) that is being used to represent state. When the qubit(s) are in a superposition multiple states can be represented at once.

5.1.5 *Quantum Circuits*

If we take a quantum state, representing one or more qubits, and apply a sequence of unitary operators (quantum gates) the result is a quantum circuit. We now take a register and let gates act on qubits, in analogy to a conventional circuit.

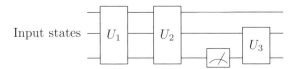

This gives us a simple form of quantum circuit (above) which is a series of operations and measurements on the state of n-qubits. Each operation is unitary and can be described by an $2^n \times 2^n$ matrix.

Each of the lines is an abstract *wire*, while the boxes containing U_n are *quantum logic gates* (or a series of gates) and the meter symbol is a measurement. Together, the gates, wires, input and output mechanisms implement quantum algorithms. Unlike classical circuits which can contain loops, quantum circuits are "one shot circuits" that just run once from left to right (and are special purpose: i.e. we have a different circuit for each algorithm).

It should be noted that it is always possible to rearrange quantum circuits so that all the measurements are done at the end of the circuit.

5.1.5.1 *Single Qubit Gates*

Just as a single qubit can be represented by a column vector, a gate acting on the qubit can be represented by a 2×2 matrix. The quantum equivalent of a NOT gate, for example, has the following form:

$$\begin{bmatrix} 0 & 1 \\ 1 & 0 \end{bmatrix}.$$

The only constraint these gates have to satisfy (as required by quantum mechanics) is that they have to be unitary, where a unitary matrix is one that satisfies the condition:

$$U^\dagger U = I.$$

This allows for a lot of potential gates.

The matrix acts as a quantum operator on a qubit. The operator's matrix must be unitary because the resultant values must satisfy the normalisation condition. Unitarity implies that the probability amplitudes must still sum to 1. If (before the gate is applied)

$$|\alpha|^2 + |\beta|^2 = 1$$

then, after the gate is applied:

$$|\alpha'|^2 + |\beta'|^2 = 1 \tag{5.23}$$

where α' and β' are the values for the probability amplitudes for the qubit after the operation has been applied.

5.1.5.2 *Pauli I Gate*

This is the identity gate.

$$\sigma_0 = I = \begin{bmatrix} 1 & 0 \\ 0 & 1 \end{bmatrix} \tag{5.24}$$

which gives us the following:

$$|0\rangle \to I \to |0\rangle, \tag{5.25}$$
$$|1\rangle \to I \to |1\rangle, \tag{5.26}$$

$$\alpha|0\rangle + \beta|1\rangle \to I \to \alpha|0\rangle + \beta|1\rangle. \tag{5.27}$$

5.1.5.3   *Pauli X Gate*

The Pauli X gate is a quantum NOT gate.

$$\sigma_1 = \sigma_X = X = \begin{bmatrix} 0 & 1 \\ 1 & 0 \end{bmatrix} \tag{5.28}$$

which gives us the following:

$$|0\rangle \to X \to |1\rangle, \tag{5.29}$$

$$|1\rangle \to X \to |0\rangle, \tag{5.30}$$

$$\alpha|0\rangle + \beta|1\rangle \to X \to \beta|0\rangle + \alpha|1\rangle. \tag{5.31}$$

The operation of the Pauli X gate can be visualised on the Bloch sphere is shown below. Here (as with any further Bloch sphere diagrams) the top point is the original state vector and the bottom point is the state vector after the transformation.

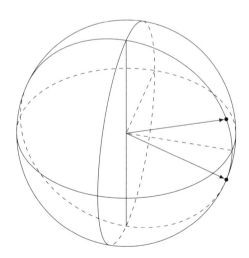

5.1.5.4 *Pauli Y Gate*

$$-\boxed{Y}-$$

$$\sigma_2 = \sigma_Y = Y = \begin{bmatrix} 0 & -i \\ i & 0 \end{bmatrix} \tag{5.32}$$

which gives us the following:

$$|0\rangle \to Y \to i|1\rangle, \tag{5.33}$$
$$|1\rangle \to Y \to -i|0\rangle, \tag{5.34}$$

$$\alpha|0\rangle + \beta|1\rangle \to Y \to -\beta i|0\rangle + \alpha i|1\rangle. \tag{5.35}$$

5.1.5.5 *Pauli Z Gate*

This gate flips a qubit's sign, i.e. changes the relative phase by a factor of -1.

$$-\boxed{Z}-$$

$$\sigma_3 = \sigma_Y = Z = \begin{bmatrix} 1 & 0 \\ 0 & -1 \end{bmatrix} \tag{5.36}$$

which gives us the following:

$$|0\rangle \to Z \to |0\rangle, \tag{5.37}$$
$$|1\rangle \to Z \to -|1\rangle, \tag{5.38}$$

$$\alpha|0\rangle + \beta|1\rangle \to Z \to \alpha|0\rangle - \beta|1\rangle. \tag{5.39}$$

5.1.5.6 *Phase Gate*

$$-\boxed{S}-$$

$$S = \begin{bmatrix} 1 & 0 \\ 0 & i \end{bmatrix} \tag{5.40}$$

which gives us the following:

$$|0\rangle \to S \to |0\rangle, \tag{5.41}$$
$$|1\rangle \to S \to i|1\rangle, \tag{5.42}$$

$$\alpha|0\rangle + \beta|1\rangle \to S \to \alpha|0\rangle + \beta i|1\rangle. \tag{5.43}$$

Note, the Phase gate can be expressed in terms of the T gate (see below):

$$S = T^2 \tag{5.44}$$

5.1.5.7 $\frac{\pi}{8}$ *Gate (T Gate)*

$$-\boxed{T}-$$

$$T = \begin{bmatrix} 1 & 0 \\ 0 & e^{i\frac{\pi}{4}} \end{bmatrix} \tag{5.45}$$

which gives us the following:

$$|0\rangle \to T \to |0\rangle, \tag{5.46}$$
$$|1\rangle \to T \to e^{i\frac{\pi}{4}}|1\rangle, \tag{5.47}$$

$$\alpha|0\rangle + \beta|1\rangle \to T \to \alpha|0\rangle + e^{i\frac{\pi}{4}}\beta|1\rangle. \tag{5.48}$$

If we apply T again we get the same as applying S once.

5.1.5.8 *Hadamard Gate*

Sometimes called the *square root of NOT gate*, it turns a $|0\rangle$ or a $|1\rangle$ into a superposition (note the different sign). This gate is one of the most important in quantum computing. We'll use this gate later for a demonstration of a simple algorithm.

$$\boxed{H}$$

$$H = \frac{1}{\sqrt{2}} \begin{bmatrix} 1 & 1 \\ 1 & -1 \end{bmatrix} \tag{5.49}$$

which gives us the following:

$$|0\rangle \to H \to \frac{1}{\sqrt{2}}(|0\rangle + |1\rangle), \tag{5.50}$$

$$|1\rangle \to H \to \frac{1}{\sqrt{2}}(|0\rangle - |1\rangle), \tag{5.51}$$

$$\alpha|0\rangle + \beta|1\rangle \to H \to \alpha\left(\frac{|0\rangle + |1\rangle}{\sqrt{2}}\right) + \beta\left(\frac{|0\rangle - |1\rangle}{\sqrt{2}}\right). \tag{5.52}$$

Example   Using H and Z gates and measuring in the $\{|+\rangle, |-\rangle\}$ basis.

(1) We can put $|0\rangle$ into state $|+\rangle$ by using an H gate:

$$|0\rangle \to H \to \frac{1}{\sqrt{2}}(|0\rangle + |1\rangle).$$

(2) We can put $|0\rangle$ into state $|-\rangle$ by using an H gate followed by a Z gate:

$$|0\rangle \to H \to \frac{1}{\sqrt{2}}(|0\rangle + |1\rangle) \to Z \to \frac{1}{\sqrt{2}}(|0\rangle - |1\rangle).$$

The operation of the H gate can be visualised on the Bloch sphere as follows:

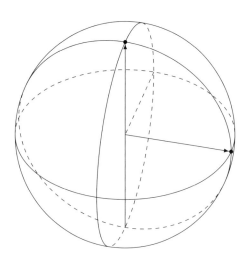

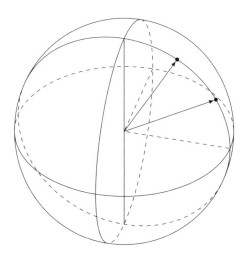

5.1.5.9 *Outer Product Notation*

A handy way to represent gates is with outer product notation. For example, a Pauli X gate can be represented by:

$$|1\rangle\langle 0| + |0\rangle\langle 1|.$$

When applied to $\alpha|0\rangle + \beta|1\rangle$, we get:

$$
\begin{aligned}
X(\alpha|0\rangle + \beta|1\rangle) &= (|1\rangle\langle 0| + |0\rangle\langle 1|)(\alpha|0\rangle + \beta|1\rangle) \\
&= |1\rangle\langle 0|(\alpha|0\rangle + \beta|1\rangle) + |0\rangle\langle 1|(\alpha|0\rangle + \beta|1\rangle) \\
&= \alpha|1\rangle 1 + \beta|1\rangle 0 + \alpha|0\rangle 0 + \beta|0\rangle 1 \\
&= \beta|0\rangle + \alpha|1\rangle.
\end{aligned}
$$

For the above it's useful to remember the following:

$$
\begin{aligned}
\langle 0|0\rangle &= 1, \\
\langle 0|1\rangle &= 0, \\
\langle 1|0\rangle &= 0, \\
\langle 1|1\rangle &= 1.
\end{aligned}
$$

Instead of doing all that mathematics, just think of it this way. For each component of the sequence, take the bra part, $\langle u|$ from $|v\rangle\langle u|$, and the new qubit's coefficient will be the old qubit's coefficient for the ket part $|v\rangle$.

Example Say we use this method on the Pauli Y outer product representation,

$$i|1\rangle\langle 0| - i|0\rangle\langle 1|.$$

When applied to $|\Psi\rangle = \alpha|0\rangle + \beta|1\rangle$ we'll see what we get:

The first part of the outer product notation is $i|1\rangle\langle 0|$. This means we take the $\alpha|0\rangle$ part of $|\Psi\rangle$ and convert it to $i\alpha|1\rangle$, so our partially built state now looks like:

$$|\Psi\rangle = \ldots + i\alpha|1\rangle.$$

Now we take the second part, $-i|0\rangle\langle 1|$. $\beta|1\rangle$ becomes $-i\beta|0\rangle$ and finally we get:

$$|\Psi\rangle = -i\beta|0\rangle + i\alpha|1\rangle.$$

Finally, the coefficients of outer product representations are the same as the matrix entries, so for the matrix:

$$\begin{bmatrix} \alpha_{00} & \alpha_{01} \\ \alpha_{10} & \alpha_{11} \end{bmatrix}.$$

The outer product representation looks like:

$$\alpha_{00}|0\rangle\langle 0| + \alpha_{01}|0\rangle\langle 1| + \alpha_{10}|1\rangle\langle 0| + \alpha_{11}|1\rangle\langle 1|.$$

5.1.5.10 *Further Properties of the Pauli Gates*

Next we'll look at the eigenvectors, eigenvalues, spectral decomposition, and outer product representation of the Pauli gates.

I has eigenvectors $|0\rangle$ and $|1\rangle$ with eigenvalues of 1 and 1 respectively. Using the spectral decomposition theorem:

$$I = 1 \cdot |0\rangle\langle 0| + 1 \cdot |1\rangle\langle 1|$$
$$= |0\rangle\langle 0| + |1\rangle\langle 1|. \tag{5.53}$$

X has eigenvectors $\frac{1}{\sqrt{2}}(|0\rangle + |1\rangle)$, and $\frac{1}{\sqrt{2}}(|0\rangle - |1\rangle)$ with eigenvalues of 1 and -1 respectively.

$$X = 1 \cdot \frac{1}{\sqrt{2}}(|0\rangle + |1\rangle)\frac{1}{\sqrt{2}}(\langle 0| + \langle 1|) + (-1) \cdot \frac{1}{\sqrt{2}}(|0\rangle - |1\rangle)\frac{1}{\sqrt{2}}(\langle 0| - \langle 1|)$$
$$= |1\rangle\langle 0| + |0\rangle\langle 1|. \tag{5.54}$$

Y has eigenvectors $\frac{1}{\sqrt{2}}(-i|0\rangle + |1\rangle)$, and $\frac{1}{\sqrt{2}}(|0\rangle - i|1\rangle)$ with eigenvalues of 1 and -1 respectively.

$$Y = 1 \cdot \frac{1}{\sqrt{2}}(-i|0\rangle + |1\rangle)\frac{1}{\sqrt{2}}(i\langle 0| + \langle 1|) + (-1) \cdot \frac{1}{\sqrt{2}}(|0\rangle - i|1\rangle)\frac{1}{\sqrt{2}}(\langle 0| + i\langle 1|)$$
$$= i|1\rangle\langle 0| - i|0\rangle\langle 1|. \tag{5.55}$$

Z has eigenvectors $|0\rangle$, and $|1\rangle$ with eigenvalues of 1 and -1 respectively.

$$Z = 1 \cdot |0\rangle\langle 0| + (-1) \cdot |1\rangle\langle 1|$$
$$= |0\rangle\langle 0| - |1\rangle\langle 1|. \tag{5.56}$$

The Pauli matrices are:

Unitary $\sigma_k \sigma_k^\dagger = I \ \forall \ k.$ $\tag{5.57}$

Hermitian $\sigma_k^\dagger = \sigma_k \ \forall \ k.$ $\tag{5.58}$

5.1.5.11 *Rotation Operators*

There are three useful operators that work well with the Bloch sphere. These are the *rotation operators* R_X, R_Y, and R_Z.

$$R_X = \begin{bmatrix} \cos\frac{\theta}{2} & -i\sin\frac{\theta}{2} \\ -i\sin\frac{\theta}{2} & \cos\frac{\theta}{2} \end{bmatrix} \tag{5.59}$$

$$= e^{-i\theta X/2}, \tag{5.60}$$

$$R_Y = \begin{bmatrix} \cos\frac{\theta}{2} & -\sin\frac{\theta}{2} \\ \sin\frac{\theta}{2} & \cos\frac{\theta}{2} \end{bmatrix} \tag{5.61}$$

$$= e^{-i\theta Y/2}, \tag{5.62}$$

$$R_Z = \begin{bmatrix} e^{-i\theta/2} & 0 \\ 0 & -e^{-i\theta/2} \end{bmatrix} \tag{5.63}$$

$$= e^{-i\theta Z/2}. \tag{5.64}$$

The rotation operators can be rewritten as

$$\cos\frac{\theta}{2}I - i\sin\frac{\theta}{2}P_\sigma \tag{5.65}$$

where P_σ means a Pauli operator identified by $\sigma = X, Y$, or Z.

In fact if we assume different angles for θ then all single qubit gates can be represented by the product of R_Y and R_Z.

Example We can represent $R_Y(90°)$ by the following matrix:

$$\frac{1}{\sqrt{2}}\begin{bmatrix} 1 & -1 \\ 1 & 1 \end{bmatrix} = \begin{bmatrix} \frac{1}{\sqrt{2}} & -\frac{1}{\sqrt{2}} \\ \frac{1}{\sqrt{2}} & \frac{1}{\sqrt{2}} \end{bmatrix}.$$

So if we apply the gate to state $|\Psi\rangle = |1\rangle$ we get the following:

$$\begin{bmatrix} \frac{1}{\sqrt{2}} & -\frac{1}{\sqrt{2}} \\ \frac{1}{\sqrt{2}} & \frac{1}{\sqrt{2}} \end{bmatrix}\begin{bmatrix} 0 \\ 1 \end{bmatrix} = \begin{bmatrix} -\frac{1}{\sqrt{2}} \\ \frac{1}{\sqrt{2}} \end{bmatrix}$$

$$= -\frac{1}{\sqrt{2}}|0\rangle + \frac{1}{\sqrt{2}}|1\rangle$$

$$= \frac{1}{\sqrt{2}}|0\rangle - \frac{1}{\sqrt{2}}|1\rangle.$$

At that last step we multiplied the entire state by a global phase factor of -1.

5.1.5.12 *Multi Qubit Gates*

A true quantum gate must be reversible. This requires that multi qubit gates use a control line, where the control line is unaffected by the unitary transformation. We'll look again at the reversible gates that were introduced in chapter 2, this time with emphasis on quantum computing.

In the case of the CNOT gate, the $\oplus$ is a classical XOR with the input on the b line and the control line a. Because it is a two qubit gate it is represented by a 4×4 matrix:

$$\begin{bmatrix} 1 & 0 & 0 & 0 \\ 0 & 1 & 0 & 0 \\ 0 & 0 & 0 & 1 \\ 0 & 0 & 1 & 0 \end{bmatrix} \tag{5.66}$$

which gives the following:

$$|00\rangle \rightarrow \text{CNOT} \rightarrow |00\rangle, \tag{5.67}$$

$$|01\rangle \rightarrow \text{CNOT} \rightarrow |01\rangle, \tag{5.68}$$

$$|10\rangle \rightarrow \text{CNOT} \rightarrow |11\rangle, \tag{5.69}$$

$$|11\rangle \rightarrow \text{CNOT} \rightarrow |10\rangle, \tag{5.70}$$

$$(\alpha|0\rangle + \beta|1\rangle)|1\rangle \rightarrow \text{CNOT} \rightarrow \alpha|01\rangle + \beta|10\rangle, \tag{5.71}$$

$$|0\rangle(\alpha|0\rangle + \beta|1\rangle) \rightarrow \text{CNOT} \rightarrow \alpha|00\rangle + \beta|01\rangle, \tag{5.72}$$

$$|1\rangle(\alpha|0\rangle + \beta|1\rangle) \rightarrow \text{CNOT} \rightarrow \alpha|11\rangle + \beta|10\rangle. \tag{5.73}$$

Example Evaluating $(\alpha|0\rangle + \beta|1\rangle)|0\rangle \rightarrow \text{CNOT} \rightarrow \alpha|00\rangle + \beta|11\rangle$.

$(\alpha|0\rangle + \beta|1\rangle)|0\rangle$ expanded is $\alpha|00\rangle + \beta|10\rangle$, so in matrix form we have:

$$\begin{bmatrix} 1 & 0 & 0 & 0 \\ 0 & 1 & 0 & 0 \\ 0 & 0 & 0 & 1 \\ 0 & 0 & 1 & 0 \end{bmatrix} \begin{bmatrix} \alpha \\ 0 \\ \beta \\ 0 \end{bmatrix} = \begin{bmatrix} \alpha \\ 0 \\ 0 \\ \beta \end{bmatrix}$$

$$= \alpha|00\rangle + \beta|11\rangle.$$

5.1.5.13 *Qubit Two NOT Gate*

As distinct from the CNOT gate we have a NOT$_2$ gate, which just does NOT on qubit two and has the following matrix representation:

$$\begin{bmatrix} 0 & 1 & 0 & 0 \\ 1 & 0 & 0 & 0 \\ 0 & 0 & 0 & 1 \\ 0 & 0 & 1 & 0 \end{bmatrix} \tag{5.74}$$

which gives the following:

$$|00\rangle \rightarrow \text{NOT}_2 \rightarrow |01\rangle, \tag{5.75}$$

$$|01\rangle \rightarrow \text{NOT}_2 \rightarrow |00\rangle, \tag{5.76}$$

$$|10\rangle \rightarrow \text{NOT}_2 \rightarrow |11\rangle, \tag{5.77}$$

$$|11\rangle \rightarrow \text{NOT}_2 \rightarrow |10\rangle. \tag{5.78}$$

Although it's not a commonly used gate it's interesting to note that the gate can be represented as a Kronecker product of I and X as follows [25]:

$$\text{NOT}_2 = I \otimes X \tag{5.79}$$

$$= \begin{bmatrix} 1 & 0 \\ 0 & 1 \end{bmatrix} \otimes \begin{bmatrix} 0 & 1 \\ 1 & 0 \end{bmatrix} \tag{5.80}$$

$$= \begin{pmatrix} 1 \begin{bmatrix} 0 & 1 \\ 1 & 0 \end{bmatrix} & 0 \begin{bmatrix} 0 & 1 \\ 1 & 0 \end{bmatrix} \\ 0 \begin{bmatrix} 0 & 1 \\ 1 & 0 \end{bmatrix} & 1 \begin{bmatrix} 0 & 1 \\ 1 & 0 \end{bmatrix} \end{pmatrix}. \tag{5.81}$$

So, as well as using the NOT_2 notation we can use the tensor product of Pauli gates on qubits one and two, shown below:

$$|00\rangle \rightarrow I \otimes X \rightarrow |01\rangle, \tag{5.82}$$

$$|01\rangle \rightarrow I \otimes X \rightarrow |00\rangle, \tag{5.83}$$

$$|10\rangle \rightarrow I \otimes X \rightarrow |11\rangle, \tag{5.84}$$

$$|11\rangle \rightarrow I \otimes X \rightarrow |10\rangle. \tag{5.85}$$

5.1.5.14   *Toffoli Gate*

The Toffoli gate was first introduced in chapter 2. Here we'll look at some properties that relate to quantum computing. The most important property is that any classical circuit can be emulated by using Toffoli gates.

The Toffoli gate can simulate NAND gates and it can perform FANOUT, which is classical bit copying (but we can't copy superposed probability amplitudes). FANOUT is easy in classical computing but impossible in quantum computing because of the no-cloning theorem (see chapter 6).

A Toffoli gate can be simulated using a number of $H, T,$ and S gates.

5.1.5.15   *Fredkin Gate*

Also introduced in chapter 2, the Fredkin gate is another three qubit gate. This gate can simulate AND, NOT, CROSSOVER, and FANOUT, it also has the interesting property that it conserves 1's.

5.2 Important Properties of Quantum Circuits

Quantum circuit diagrams have the following constraints which make them different from classical diagrams.

(1) They are acyclic (no loops).
(2) No FANIN, as FANIN implies that the circuit is NOT reversible, and therefore not unitary.
(3) No FANOUT, as we cannot copy a qubit's state during the computational phase because of the no-cloning theorem (explained in chapter 6).

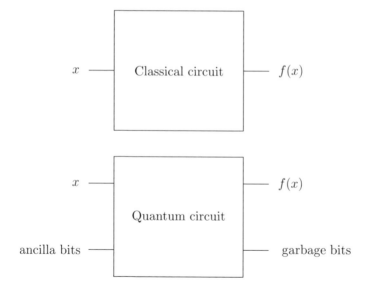

Fig. 5.2   Garbage and ancilla bits.

All of the above can be simulated with the use of ancilla and garbage bits if we assume that no qubits will be in a superposition (figure 5.2). As stated in chapter 2, garbage bits are useless qubits left over after computation and ancilla bits are extra qubits needed for temporary calculations.

5.2.1 *Common Circuits*

5.2.1.1 *Controlled U Gate*

Let U be a unitary matrix which uses an arbitrary number of qubits. A controlled U gate is a U gate with a control line, i.e. if the control qubit is $|1\rangle$ then U acts on its data qubits, otherwise they are left alone.

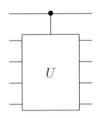

5.2.1.2 *Bit Swap Circuit*

This circuit swaps the values of qubits between lines.

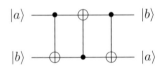

The circuit can be simplified to:

$$|a\rangle \quad \text{---}\ast\text{---}\quad |b\rangle$$
$$|b\rangle \quad \text{---}\ast\text{---}\quad |a\rangle$$

Here are some examples:

$$|0\rangle \quad \text{---}\ast\text{---}\quad |1\rangle$$
$$|1\rangle \quad \text{---}\ast\text{---}\quad |0\rangle$$

$$|1\rangle \quad \text{---}\ast\text{---}\quad |0\rangle$$
$$|0\rangle \quad \text{---}\ast\text{---}\quad |1\rangle$$

5.2.1.3 *Copying Circuit*

If we have no superposition then we can copy bits in a classical sense with a CNOT.

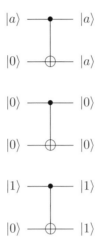

But when we use a superposition as input:

$$\alpha\,|0\rangle + \beta\,|1\rangle \quad \bullet$$
$$|0\rangle \quad \oplus$$

The combined state becomes:

$$(\alpha|0\rangle + \beta|1\rangle)|0\rangle = \alpha|00\rangle + \beta|10\rangle,$$

$$\alpha|00\rangle + \beta|10\rangle \rightarrow \text{CNOT} \rightarrow \alpha|00\rangle + \beta|11\rangle.$$

Which is **not** a copy of the original state because:

$$(\alpha|0\rangle + \beta|1\rangle)(\alpha|0\rangle + \beta|1\rangle) \neq \alpha|00\rangle + \beta|11\rangle.$$

A qubit in an unknown state (as an input) cannot be copied. When it is copied it must first be measured to be copied. The information held in the probability amplitudes α and β is lost.

5.2.1.4 *Bell State Circuit*

This circuit produces Bell states that are entangled. We'll represent a Bell state circuit by β, and the individual Bell states as $|\beta_{00}\rangle, |\beta_{01}\rangle, |\beta_{10}\rangle$ and $|\beta_{11}\rangle$.

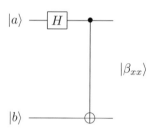

$$|00\rangle \rightarrow \beta \rightarrow \frac{1}{\sqrt{2}}(|00\rangle + |11\rangle) = |\beta_{00}\rangle,$$

$$|01\rangle \rightarrow \beta \rightarrow \frac{1}{\sqrt{2}}(|01\rangle + |10\rangle) = |\beta_{01}\rangle,$$

$$|10\rangle \rightarrow \beta \rightarrow \frac{1}{\sqrt{2}}(|00\rangle - |11\rangle) = |\beta_{10}\rangle,$$

$$|11\rangle \rightarrow \beta \rightarrow \frac{1}{\sqrt{2}}(|01\rangle - |10\rangle) = |\beta_{11}\rangle.$$

5.2.1.5 *Superdense Coding*

It's possible, by using entangled pairs, to communicate two bits of information by transmitting one qubit. Here's how:

(1) Initially Alice and Bob each take one half of an EPR pair, say we start with β_{00}. This means that Alice has the **bolded** qubit in $\frac{1}{\sqrt{2}}(|\mathbf{0}0\rangle + |\mathbf{1}1\rangle)$ and Bob has the **bolded** qubit in $\frac{1}{\sqrt{2}}(|0\mathbf{0}\rangle + |1\mathbf{1}\rangle)$. They then move apart to an arbitrary distance.

(2) Depending on which value Alice wants to send to Bob, she applies a gate (or gates) to her qubit. This is described below and the *combined* state

is shown after the gate's operation with Alice's qubit shown **bolded**:

$$|\mathbf{0}0\rangle \to I \to \frac{1}{\sqrt{2}}(|\mathbf{0}0\rangle + |\mathbf{1}1\rangle),$$

$$|\mathbf{1}0\rangle \to X(X|0\rangle = |1\rangle, X|1\rangle = |0\rangle) \to \frac{1}{\sqrt{2}}(|\mathbf{1}0\rangle + |\mathbf{0}1\rangle),$$

$$|\mathbf{0}1\rangle \to Z(Z|0\rangle = |0\rangle, Z|1\rangle = -|1\rangle) \to \frac{1}{\sqrt{2}}(|\mathbf{0}0\rangle - |\mathbf{1}1\rangle),$$

$$|\mathbf{1}1\rangle \to XZ \to \frac{1}{\sqrt{2}}(|\mathbf{0}1\rangle - |\mathbf{1}0\rangle).$$

(3) Alice now sends her qubit to Bob via a classical channel.
(4) Bob now uses a CNOT which allows him to factor out the second qubit while the first one stays in a superposition.

$$\frac{1}{\sqrt{2}}(\alpha|00\rangle + \beta|11\rangle) \to \mathtt{CNOT} \to \frac{1}{\sqrt{2}}(\alpha|00\rangle + \beta|10\rangle) = \frac{1}{\sqrt{2}}(\alpha|0\rangle + \beta|1\rangle)|0\rangle,$$

$$\frac{1}{\sqrt{2}}(\alpha|10\rangle + \beta|01\rangle) \to \mathtt{CNOT} \to \frac{1}{\sqrt{2}}(\alpha|01\rangle + \beta|11\rangle) = \frac{1}{\sqrt{2}}(\alpha|0\rangle + \beta|1\rangle)|1\rangle,$$

$$\frac{1}{\sqrt{2}}(\alpha|00\rangle - \beta|11\rangle) \to \mathtt{CNOT} \to \frac{1}{\sqrt{2}}(\alpha|00\rangle - \beta|10\rangle) = \frac{1}{\sqrt{2}}(\alpha|0\rangle - \beta|1\rangle)|0\rangle,$$

$$\frac{1}{\sqrt{2}}(\alpha|01\rangle - \beta|10\rangle) \to \mathtt{CNOT} \to \frac{1}{\sqrt{2}}(\alpha|01\rangle - \beta|11\rangle) = \frac{1}{\sqrt{2}}(\alpha|0\rangle - \beta|1\rangle)|1\rangle.$$

(5) Now Bob applies an H gate to the first bit to collapse the superposition.

$$\frac{1}{\sqrt{2}}(\alpha|0\rangle - \beta|1\rangle) \to H \to |1\rangle,$$

$$\frac{1}{\sqrt{2}}(\alpha|0\rangle + \beta|1\rangle) \to H \to |0\rangle.$$

So Bob gets the following:

$$\frac{1}{\sqrt{2}}(\alpha|0\rangle + \beta|1\rangle)|0\rangle \to (H \otimes I) \to |00\rangle,$$

$$\frac{1}{\sqrt{2}}(\alpha|0\rangle + \beta|1\rangle)|1\rangle \to (H \otimes I) \to |01\rangle,$$

$$\frac{1}{\sqrt{2}}(\alpha|0\rangle - \beta|1\rangle)|0\rangle \to (H \otimes I) \to |10\rangle,$$

$$\frac{1}{\sqrt{2}}(\alpha|0\rangle - \beta|1\rangle)|1\rangle \to (H \otimes I) \to |11\rangle.$$

Bob can now measure the two qubits in the computational basis and the result will be the value that Alice wanted to send.

5.2.1.6   *Teleportation Circuit*

Teleportation is basically the opposite of superdense coding, i.e. superdense coding takes a quantum state to two classical bits. Teleportation takes two classical bits to one quantum state.

Alice's circuit

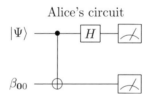

Bob chooses one of the following four circuits

$$\beta_{00} \;-\!\boxed{I}\!-\; |\Psi\rangle$$

$$\beta_{00} \;-\!\boxed{X}\!-\; |\Psi\rangle$$

$$\beta_{00} \;-\!\boxed{Z}\!-\; |\Psi\rangle$$

$$\beta_{00} \;-\!\boxed{X}\!-\!\boxed{Z}\!-\; |\Psi\rangle$$

(1) Like superdense coding, initially Alice and Bob each take one half of an EPR pair, say we start with β_{00}. This means that Alice has the

bolded qubit in $\frac{1}{\sqrt{2}}(|\mathbf{00}\rangle + |\mathbf{11}\rangle)$ and Bob has the **bolded** qubit in $\frac{1}{\sqrt{2}}(|00\rangle + |11\rangle)$. They then move apart to an arbitrary distance.

(2) Alice has a qubit in an unknown state:

$$|\Psi\rangle = \alpha|0\rangle + \beta|1\rangle.$$

Which she combines with her entangled qubit:

$$(\alpha|0\rangle + \beta|1\rangle)\left(\frac{1}{\sqrt{2}}(|00\rangle + |11\rangle)\right).$$

This gives the following combined state (Alice's qubits are bolded):

$$|\Psi\rangle = \frac{1}{\sqrt{2}}(\alpha|\mathbf{00}0\rangle + \alpha|\mathbf{01}1\rangle + \beta|\mathbf{10}0\rangle + \beta|\mathbf{11}1\rangle).$$

(3) Alice then applies a CNOT. Note — this is like using (CNOT $\otimes I$) on the *combined* three qubit system, i.e. including Bob's qubit.

$$\frac{1}{\sqrt{2}}(\alpha|\mathbf{00}0\rangle + \alpha|\mathbf{01}1\rangle + \beta|\mathbf{11}0\rangle + \beta|\mathbf{10}1\rangle).$$

(4) Alice then applies an H gate to her first qubit, the qubit we want to teleport (or $H \otimes I \otimes I$ for the combined system):

$$\frac{1}{2}(\alpha(|000\rangle + |100\rangle + |011\rangle + |111\rangle) + \beta(|010\rangle - |110\rangle + |001\rangle - |101\rangle)).$$

Now we rearrange the state to move amplitudes α and β so that we can read the first two bits leaving the third in a superposition.

$$=\frac{1}{2}(|00\rangle\alpha|0\rangle + |10\rangle\alpha|0\rangle + |01\rangle\alpha|1\rangle + |11\rangle\alpha|1\rangle + |01\rangle\beta|0\rangle - |11\rangle\beta|0\rangle$$
$$+ |00\rangle\beta|1\rangle - |10\rangle\beta|1\rangle),$$
$$=\frac{1}{2}(|\mathbf{00}\rangle(\alpha|0\rangle + \beta|1\rangle) + |\mathbf{01}\rangle(\alpha|1\rangle + \beta|0\rangle) + |\mathbf{10}\rangle(\alpha|0\rangle - \beta|1\rangle)$$
$$+ |\mathbf{11}\rangle(\alpha|1\rangle - \beta|0\rangle)).$$

(5) Alice now performs measurements on her state to determine which of the above bolded states her qubits are in. She then communicates via a classical channel what she measured (i.e. a $|00\rangle, |01\rangle, |10\rangle$, or a $|11\rangle$) to Bob.

(6) Bob now may use X and/or Z gate(s) to fix up the phase and order of the probability amplitudes, (he selects gates based on what Alice tells him) so that the result restores the original qubit. Summarised below are the gates he must use:

$$\text{Case } \mathbf{00} = \alpha|0\rangle + \beta|1\rangle \rightarrow I \rightarrow \alpha|0\rangle + \beta|1\rangle,$$

$$\text{Case } \mathbf{01} = \alpha|1\rangle + \beta|0\rangle \rightarrow X \rightarrow \alpha|0\rangle + \beta|1\rangle,$$

$$\text{Case } \mathbf{10} = \alpha|0\rangle - \beta|1\rangle \rightarrow Z \rightarrow \alpha|0\rangle + \beta|1\rangle,$$

$$\text{Case } \mathbf{11} = \alpha|1\rangle - \beta|0\rangle \rightarrow XZ \rightarrow \alpha|0\rangle + \beta|1\rangle.$$

5.3   The Reality of Building Circuits

There is a general theorem: any unitary operation on n qubits can be implemented using a set of two qubit operations. This can include CNOTs and other single bit operations. This result resembles the classical result that any boolean function can be implemented with NAND gates. This is helpful because sometimes we are limited in what we can use to build a quantum circuit.

5.3.1   *Building a Programmable Quantum Computer*

5.3.1.1   *Is it Possible?*

Can we build a programmable quantum computer, i.e. a quantum computer that has an architecture similar to Von Neumann (or Harvard) architecture? **No!** This is because:

Distinct unitary operators $U_0...U_n$ require orthogonal programs $|U_0\rangle, ...|U_n\rangle$ [30]

This is called the *no programming theorem*.

If we were to have a programmable quantum computer our "program" would consist of one or more unitary operators. Since there are an infinite number of these unitary operators the program register would have to be infinite in size (that is our input to the quantum computer that contains the program) [30].

5.4 The Four Postulates of Quantum Mechanics

The theory of quantum mechanics has four main postulates. These are introduced here as simple sentences and then explained in more details in the following sections.

(1) In a closed quantum system we need a way of describing the state of all the particles within it. The first postulate gives us a way to do this by using a single state vector to represent the entire system. Say the state is to be a vector in $\mathbb{C}^n$, this would be $\mathbb{C}^2$ for a spin system.

(2) The evolution of a closed system is a unitary transform. While the system is evolving under its own steam — no measurement — the state at some stage $|\Psi'\rangle$ is related to the state at some previous stage (or time) $|\Psi\rangle$ by a unitary transform $|\Psi'\rangle = U|\Psi\rangle$. This means that we can fully describe the behaviour of a system by using unitary matrices.

(3) The third postulate relates to making measurements on a closed quantum system, and the affect those measurements have on that system.

(4) Postulate four relates to combining or separating different closed quantum systems using tensor products.

5.4.1 *Postulate One*

An isolated system has an associated complex vector space called a state space. We will use a state space called a *Hilbert space*.

The state of the quantum system can be described by a unit vector in this space called a state vector.

Example   The simplest system we are interested in is a qubit which is in $\mathbb{C}^2$. A qubit is a unit vector $|\Psi\rangle$ in $\mathbb{C}$. Most of the time we'll attach an orthonormal basis (like $\{|0\rangle, |1\rangle\}$). Our qubit can be described by:

$$|\Psi\rangle = \alpha|0\rangle + \beta|1\rangle = \begin{bmatrix} \alpha \\ \beta \end{bmatrix}$$

here α and β are known as probability amplitudes and we say the qubit is in a quantum superposition of states $|0\rangle$ and $|1\rangle$.

5.4.2   *Postulate Two*

5.4.2.1   *Simple Form*

An isolated (closed) system's evolution can be described by a unitary transform.

$$|\Psi'\rangle = U|\Psi\rangle. \tag{5.86}$$

5.4.2.2   *Form Including Time*

If we include time (as quantum interactions happen in continuous time):

$$|\Psi(t_2)\rangle = U(t_1, t_2)|\Psi(t_1)\rangle \tag{5.87}$$

here t_1 and t_2 are points in time and $U(t_1, t_2)$ is a unitary operator than can vary with time.

We can also say that the process is reversible, because:

$$U^\dagger U|\Psi\rangle = |\Psi\rangle. \tag{5.88}$$

The history of the quantum system does not matter as it is completely described by the current state (this is know as a *Markov process*).

Note: we can rewrite the above in terms of Schrödinger's equation, but that is beyond the scope of this tutorial.

5.4.3   *Postulate Three*

5.4.3.1   *Simple Form*

This deals with what happens if $|\Psi\rangle$ is measured in an orthonormal basis:

$$\{|O_1\rangle, |O_1\rangle, \ldots, |O_n\rangle\}.$$

We'll measure a particular outcome j with probability:

$$p_j = |\langle O_j|\Psi\rangle|^2. \tag{5.89}$$

Example   If we measure in the computational basis we have:

$$p_0 = |\langle 0|\Psi\rangle|^2$$
$$= |\langle 0|(\alpha|0\rangle + \beta|1\rangle)|^2$$
$$= |\alpha|^2$$
$$p_1 = |\beta|^2.$$

After the measurement the system is in state $|O_j\rangle$. This is because measurement disturbs the system.

If $|u_1\rangle$ and $|u_2\rangle$ are not orthogonal (i.e. $\langle u_1|u_2\rangle \neq 0$) then we can't reliably distinguish between them on measurement.

5.4.3.2   *Projectors*

Suppose we have a larger quantum system which is a combination of smaller systems, i.e. we have an orthonormal basis $A = \{|e_1\rangle, \ldots, |e_n\rangle\}$ where n is the dimension of A. Then we have a larger quantum system B such that $A \subset B$. If we measure A what is the effect on B? We can say that $|O_1\rangle, \ldots, |O_n\rangle$ is part of or comprised of one or many of the orthogonal subspaces $V_1, V_2, \ldots, V_m$ which are connected in the following way:

$$V = V_1 \oplus V_2 \oplus \ldots \oplus V_m. \tag{5.90}$$

Example The state vector:

$$|\Psi\rangle = (\alpha|O_1\rangle + \beta|O_2\rangle) + \gamma|O_3\rangle$$

can be rewritten as:

$$V(|O_1\rangle, |O_2\rangle, |O_3\rangle) = V_1(|e_1\rangle, |e_2\rangle) \oplus V_2(|e_3\rangle).$$

We can use projectors $(P_1, \ldots, P_m)$ to filter out everything other than the subspace we are looking for (i.e. everything orthogonal to our subspace V).

Example From the last example, projector P_1 on $V_1(|e_1\rangle, |e_2\rangle)$ gives us:

$$P_1(\alpha|e_1\rangle + \beta|e_2\rangle + \gamma|e_3\rangle) = \alpha|e_1\rangle + \beta|e_2\rangle.$$

Formally, if $P_1, \ldots, P_m$ is a set of projectors which covers all the orthogonal subspaces of the state space, upon measuring $|\Psi\rangle$ we have,

$$p_j = \langle\Psi|P_j|\Psi\rangle \tag{5.91}$$

leaving the system in the post measurement state:

$$\frac{P_j|\Psi\rangle}{\sqrt{\langle\Psi|P_j|\Psi\rangle}} \, . \tag{5.92}$$

Example Given a qutrit $|\Psi\rangle = \alpha|0\rangle + \beta|1\rangle + \gamma|2\rangle)$.

P_1 on $V_1(|0\rangle, |1\rangle)$ and P_2 on $V_2(|2\rangle)$ gives us:

$$p_1 = \langle\Psi|P_1|\Psi\rangle$$

$$= [\alpha^* \beta^* \gamma^*] \begin{bmatrix} \alpha \\ \beta \\ 0 \end{bmatrix}$$

$$= |\alpha|^2 + |\beta|^2,$$

$$p_2 = \langle\Psi|P_2|\Psi\rangle$$

$$= [\alpha^* \beta^* \gamma^*] \begin{bmatrix} 0 \\ 0 \\ \gamma \end{bmatrix}$$

$$= |\gamma|^2.$$

So our separated look like this:

$$|\Psi_1\rangle = \frac{P_1|\Psi\rangle}{\sqrt{\langle\Psi|P_1|\Psi\rangle}}$$

$$= \frac{\alpha|0\rangle + \beta|1\rangle}{\sqrt{|\alpha|^2 + |\beta|^2}} \, ,$$

$$|\Psi_2\rangle = \frac{P_2|\Psi\rangle}{\sqrt{\langle\Psi|P_2|\Psi\rangle}}$$

$$= \frac{\gamma|2\rangle}{\sqrt{|\gamma|^2}} \, .$$

More importantly we can look at partial measurement of a group of qubits. The following example uses the tensor product which is also part of postulate 4. If the system to be measured is in the basis $\{|O_1\rangle, |O_2\rangle\}$ then we use the projector with a tensor product with I on the qubit we do not want to measure. e.g. for $|O_2\rangle$ of qubit two we use $(I \otimes P_2)$.

Example If $|\Psi\rangle = \alpha_{00}|00\rangle + \alpha_{01}|01\rangle + \alpha_{10}|10\rangle + \alpha_{11}|11\rangle$:

For measuring qubit one in the computational basis we get $|0\rangle$ with probability:

$$
\begin{aligned}
p_0 &= \langle \Psi | P_0 \otimes I | \Psi \rangle \\
&= (\alpha_{00}|00\rangle + \alpha_{01}|01\rangle + \alpha_{10}|10\rangle + \alpha_{11}|11\rangle) \bullet (\alpha_{00}|00\rangle + \alpha_{01}|01\rangle) \\
&= |\alpha_{00}|^2 + |\alpha_{01}|^2.
\end{aligned}
$$

There is another type of measurement called a *POVM* (Positive Operator Valued Measures) of which projectors are a certain type. POVMs are beyond the scope of this text.

5.4.4 *Postulate Four*

A tensor product of the components of a composite physical system, describes the system. The state spaces of the individual systems are combined so:

$$
\mathbb{C}^n \otimes \mathbb{C}^n = \mathbb{C}^{n^2}. \tag{5.93}
$$

An example is on the next page.

Example $\mathbb{C}^4 \otimes \mathbb{C}^4 = \mathbb{C}^{16}$ can look like:

$$(|\Psi_A\rangle = |1\rangle + |2\rangle + |3\rangle + |4\rangle) \otimes (|\Psi_B\rangle = |a\rangle + |b\rangle + |c\rangle + |d\rangle)$$

and can be written as:

$$|\Psi_{AB}\rangle = |1a\rangle + |1b\rangle + |1c\rangle + \ldots + |4d\rangle.$$

Example If Alice has $|\Psi_A\rangle = |u\rangle$ and Bob has $|\Psi_B\rangle = |v\rangle$, then if their systems are combined the joint state is:

$$|\Psi_{AB}\rangle = |u\rangle \otimes |v\rangle.$$

If Bob applies gate U to this system it means $I \otimes U$ is applied to the joint system.

Example Given the following:

$$|\Psi\rangle = \sqrt{0.1}|00\rangle + \sqrt{0.2}|01\rangle + \sqrt{0.3}|10\rangle + \sqrt{0.4}|11\rangle$$

then,

$$|\Psi\rangle \rightarrow (I \otimes X) \rightarrow |\Psi\rangle = \sqrt{0.1}|01\rangle + \sqrt{0.2}|00\rangle + \sqrt{0.3}|11\rangle + \sqrt{0.4}|10\rangle,$$

$$|\Psi\rangle \rightarrow (X \otimes I) \rightarrow |\Psi\rangle = \sqrt{0.1}|10\rangle + \sqrt{0.2}|11\rangle + \sqrt{0.3}|00\rangle + \sqrt{0.4}|01\rangle.$$

Chapter 6

Information Theory

6.1 Introduction

Information theory examines the ways information can be represented and transformed efficiently [37]. This information can be represented in many different ways to express the same meaning. E.g. "How are you?" and "Comment allez vous?" express the same meaning. All known ways of representing this information must have a physical medium like magnetic storage or ink on paper. The information stored on a particular physical medium is not tied to that medium and can be converted from one form to another. If we consider information in these terms then it becomes a property like energy that can be transferred from one physical system to another.

We are interested in quantum information, which has many parallels in conventional information theory. Conventional information theory relies heavily on the classical theorems of Claude E. Shannon, 1916–2001 (figure 6.1). There are a number of quantum equivalents for the various parts of his classical theorems. In this chapter we'll look at these and other related topics like quantum error correction and quantum cryptology. Also, as promised, there is a fairly in-depth section on Bell states and the chapter ends with some open questions on the nature of information and alternate methods of computation.

As with chapter 5 individual references to QCQI have been dropped as they would appear too frequently.

Fig. 6.1   Claude E. Shannon and George Boole.

6.2   History

The history of information theory can be said to have started with the invention of *boolean algebra* in 1847 by George Boole, 1815–1864 (figure 6.1). Boolean algebra introduced the concept of using logical operations (like AND, OR, and NOT) on the binary number system. The next milestone was in 1948 when Shannon wrote "the mathematical theory of communication" in which he outlined the concepts of Shannon entropy (see section 6.5.1) [33]. Earlier Shannon had shown that boolean algebra could be used to represent relays, switches, and other components in electronic circuits. Shannon also defined the most basic unit of information theory — the bit (binary digit).

6.3   Shannon's Communication Model

To formally describe the process of transmitting information from a source to a destination we can use *Shannon's communication model*, which is shown in figure 6.2. The components of this are described as follows:

Source — The origin of the *message*, which itself has a formal definition (see section 6.4). The *message* is sent from the source in its raw form.

Transmitter — The transmitter encodes and may compress the *message* at which point the *message* becomes a *signal* which is transported from transmitter to receiver.

Source of Noise — The noise source can introduce random noise into the *signal*, potentially scrambling it.

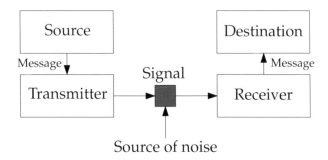

Fig. 6.2 Shannon's communication model.

Receiver — The receiver may decode and decompress the *signal* back
into the original *message*.

Destination — The destination of the raw *message*.

6.3.1 *Channel Capacity*

A message is chosen from a set of all possible messages and then transmitted. Each symbol transferred takes a certain amount of time (which is
called the *channel capacity*).

Shannon's name for channel capacity on a binary channel is "one bit
per time period" e.g. 56,000 bits per second. His expression for capacity
is:

$$C = \lim_{T \to \infty} \frac{\log_2 N}{T} \qquad (6.1)$$

where N is the number of possible messages of length T.

Example   For binary we have:

2 bits = 4 different messages in 2 time periods.
3 bits = 8 different messages in 3 time periods.

so,

$$N(T) = 2^T$$

and,

$$C = \lim_{T \to \infty} \frac{\log_2 N}{T}$$

$$= 1 \text{ bit per time period.}$$

Example   Another example is morse code, where dashes take longer to transmit than dots. If dashes are represented by 1110 and a dot is 10 for this we have:

$$C = 0.34 \text{ bit per time period.}$$

6.4   Classical Information Sources

A source of information produces a discrete set of symbols from a specific alphabet. The alphabet that is most commonly used is binary (1 and 0), but it could in principle be any series of symbols.

The way an information source can be modeled is via a probability that certain letters or combinations of letters (words) will be produced by the source. An example for this *probability distribution* would be, given an unknown book, one can, in advance, predict to a certain degree of accuracy the frequency of words and letters within the book [30].

6.4.1   *Independent Information Sources*

An *Independent and Identically Distributed* (IID) information source is an information source in which each output has no dependency on other outputs from the source, and furthermore each output has the same probability of occurring each time it is produced [30].

An IID is an information source with an alphabet (i.e a set of symbols or outputs a_i):

$$A = \{a_1, \ldots, a_n\} \tag{6.2}$$

with probabilities $p_{a_1}, p_{a_2}, \ldots, p_{a_n}$ such that:

$$\sum_{i=1}^{n} p_{a_i} = 1 \text{ where } 0 \le p_{a_i} \le 1 \; \forall \; i. \tag{6.3}$$

This source will produce a letter with probability p_{a_i} with no dependency on the previous symbols. Independent information sources are also called *zero memory information sources* (i.e. they correspond to a Markov process).

Example [30] A biased coin is a good example of an IID. The biased coin has a probability p of heads and a probability of $1-p$ of tails. Given a language $\sum = \{heads, tails\}$:

$$p_{heads} = 0.3,$$

$$p_{tails} = 0.7 \, .$$

Our coin will come up tails 70% of the time.

Strictly speaking, Shannon's results only hold for a subset of information sources that conform to the following:

(1) Symbols must be chosen with fixed probabilities, one by one with no dependency on current symbols and preceding choices.
(2) An information source must be an ergodic source. This means that there should be no statistical variation (with a probability of 1) between possible sources, i.e. all systems should have the same probabilities for letters to appear in their alphabets.

Not many sources are perfect like the above. The reason is that, for example a book has correlations between syllables, words, etc. (not just letters) like "he" and "wh" [30]. We can measure certain qualities and the source becomes more like an IID (like JUST letter frequency). Shannon suggested that most information sources can be approximated.

6.5   Classical Redundancy and Compression

Compression of data means using less information to represent a message
and reconstructing it after it has been transmitted. When we talk about
compression in this section we mean simple algorithmic compression that
can be applied to all information sources. This is distinct from say, chang-
ing the text in a sentence to convey the same meaning or using special
techniques to only work on a subset of messages or information sources
(like using a simple formula to exactly represent a picture of a sphere).

We often talk about using a coding K to represent our message. Formally, a coding is a function that takes a source alphabet A to a coding
alphabet B, i.e. $K : A \to B$. For every symbol a in the language A, $K(a)$
is a *word* with letters from B.

A word, $w = a_1 a_2 \ldots a_n$ in A is found by:

$$K(w) = K(a_1)K(a_2)\ldots K(a_n). \tag{6.4}$$

Example   A simple coding.

$$A = \{A, B, C, D\},$$
$$B = \{0, 1\}.$$

Possible encodings are:

$$A \to 0001,$$

$$B \to 0101,$$

$$C \to 1001,$$

$$D \to 1111.$$

So we encode the word $ABBA$ as:

$$K(ABBA) = 0001\ 0101\ 0101\ 0001.$$

6.5.0.1 *Length of Codes*

We define the size of an alphabet A as $|A|$ and the length of a word w as $|w|$.

Example $|A|$ and $|w|$.

$$A = \{a, b, c, d, e, f\},$$
$$B = \{0, 1, 2\}.$$

Possible encodings are:

$$a \to 0,$$
$$b \to 1,$$
$$c \to 20,$$
$$d \to 220,$$
$$e \to 221,$$
$$f \to 222.$$

We get:

$$|K(a)| = 1, |K(c)| = 2, |K(f)| = 3, |A| = 6, |B| = 3.$$

6.5.1 *Shannon's Noiseless Coding Theorem*

Shannon demonstrated that there is a definite limit to how much an information source can be compressed. *Shannon Entropy $H(S)$* is the minimum number of bits needed to convey a message. For a source S the entropy is related to the shortest average length coding $L_{min}(S)$ by:

$$H(S) \leq L_{min}(S) < H(S) + 1. \tag{6.5}$$

We can find the Shannon Entropy of a particular source distribution measured in bits by:

$$H(X) = -\sum_i p_i \log_2 p_i. \tag{6.6}$$

Here, $\log_2$ (a log of base 2) means that the Shannon entropy is measured in bits. p_i is a measure of probability (i.e. the frequency with which it is emitted) for a symbol being generated by the source and the summation is a sum over all the symbols $i = 1, 2, \ldots, n$ generated by the source.

Example Random and dependent sources.

Random Each symbol is chosen from the source totally randomly (with A and B having equal probability), with no dependencies. This source is uncompressible with Shannon Entropy of 1 bit per symbol.

Dependent Every second symbol is exactly the same as the last, which is chosen randomly (e.g. AABBBBAABB). This has Shannon Entropy of $\frac{1}{2}$ bit(s) per symbol.

Example We have a language $\{A, B, C, D, E\}$ of symbols occurring with the following frequency:

$$A = 0.5, B = 0.2, C = 0.1, D = 0.1, E = 0.1 .$$

The entropy is:

$$\begin{aligned}
H(X) &= -[(0.5 \log_2 0.5 + 0.2 \log_2 0.2 + (0.1 \log_2 0.1) \times 3)] \\
&= -[-0.5 + (-0.46438) + (-0.9965)] \\
&= -[-1.9] \\
&= 1.9.
\end{aligned}$$

So we need 2 bits per symbol to convey the message.

The minimum entropy is realised when the information source produces a single letter constantly, this gives us a probability of 1 for that letter.

The maximum entropy is realised when we have no information about the probability distribution of the source alphabet (when all symbols are equally likely to occur).

A special case of the entropy is *binary entropy* where the source has just two symbols with probabilities of p and $1 - p$ like the biased coin toss for example. Note a fair coin toss has maximum entropy of 1 bit and a totally unfair, weighted coin that always comes up heads or always comes up tails has minimum entropy of 0 bits.

6.5.2   Quantum Information Sources

Shannon entropy gives us a lower bound on how many bits we need to store a particular piece of information. The question is, is there any difference when we use quantum states? The answer is yes if we use a superposition.

   If the qubits involved are in well-defined states (like $|0\rangle$ and $|1\rangle$) a semi-classical coin toss [30] gives us:

$|0\rangle$ with probability of $\frac{1}{2}$,
$|1\rangle$ with probability of $\frac{1}{2}$,

$$H\left(\frac{1}{2}\right) = 1.$$

If we replace one of these states with a superposition then a "quantum coin toss" gives us:

$|0\rangle$ with probability of $\frac{1}{2}$,
$\frac{|0\rangle + |1\rangle}{\sqrt{2}}$ with probability of $\frac{1}{2}$,

$$H\left(\frac{1 + \frac{1}{\sqrt{2}}}{2}\right) = 0.6 \ .$$

Better than Shannon's rate!

Generally a quantum information source produces state: $|\Psi_j\rangle$ with probabilities p_j, and our quantum compression performs better than the Shannon rate $H(p_j)$.

6.5.3   Pure and Mixed States

A quantum system is said to be in a pure state if its state is well-defined. This does not mean the state vector will always collapse to a known value; but at least the state vector is known as distinct from what it collapses to. For example given a photon and a polariser the photon can be in three states: horizontal (H), vertical (V) and diagonal (D). If we make our photon diagonal at 45° then we have an equal superposition of H and V.

$$|D\rangle = \frac{1}{\sqrt{2}}|H\rangle + \frac{1}{\sqrt{2}}|V\rangle.$$

Now if we measure polarisation we have a 50% chance of detecting an $|H\rangle$ or a 50% chance of detecting a $|V\rangle$. The state was well-defined before measurement, so we call it *pure* (i.e. we have a well-defined angle of polarisation viz 45°, and if we measure with a polariser in this direction the result is certain). We don't have this situation if there is no direction in which the result is certain. So the state is not well-defined and we call this state *mixed*. For example, if we have a number of photons, 50% of which are polarised horizontally and 50% vertically, now we have a mixed state (which corresponds to a point in the interior if the Bloch Sphere). Each photon has a well defined state $|H\rangle$ or $|V\rangle$, but not the group. This situation is indistinguishable from a photon that is in an entangled state. We can call the collection of states a *mixture* or an *ensemble*.

For example, we have a mixed state containing two qubits. The first qubit is in the state below with a probability of of 2/3.

$$\frac{1}{\sqrt{2}}|0\rangle + \frac{1}{\sqrt{2}}|1\rangle$$

The second qubit is in the following state with the probability of of 1/3,

$$\frac{1}{\sqrt{2}}|0\rangle - \frac{1}{\sqrt{2}}|1\rangle$$

When closed quantum systems get affected by external systems our pure states can get entangled with the external world, leading to mixed states. This is called *decoherence*, or in information theory terms, noise.

It is possible to have a well-defined (pure) state that is composed of subsystems in mixed states. Entangled states like Bell states are well-defined for the composite system (whole), but the state of each component qubit is not well-defined, i.e. mixed.

6.5.4 *Schumacher's Quantum Noiseless Coding Theorem*

The quantum analogue of Shannon's noiseless coding theorem is *Schumacher's quantum noiseless coding theorem*. Which is as follows [30]:

The best data rate R achievable is $S(\rho)$. $\qquad\qquad$ (6.7) where S is *Von Neumann entropy* and ρ is the *density matrix*. ρ holds equivalent information to the quantum state Ψ and quantum mechanics can be formulated in terms of ρ as an alternative to Ψ. We now look at ρ.

6.5.4.1 *The Density Matrix*

In studying quantum noise it turns out to be easier to work with the density matrix than the state vector (think of it as just a tool; not a necessary component of quantum computing).

According to Nielsen [30] there are three main approaches to the density matrix: the *ensemble* point of view, *subsystem*, and *fundamental* approaches. They are described briefly here.

Ensemble — This is the basic view of the density matrix which gives us measurement statistics in a compact form.

Subsystem — If quantum system $\mathbf{A}$ is coupled to another quantum system $\mathbf{B}$ we cannot always give $\mathbf{A}$ a state vector on its own (it is not well-defined). But we can assign an individual density matrix to either subsystem.

Fundamental — It is possible to restate the four postulates of quantum mechanics in terms of the density matrix.

For example, as mentioned above, we can view the statistics generated by an entangled photon as equivalent to that of the corresponding ensemble.

6.5.4.2 *Ensemble Point of View*

Consider a collection of identical quantum systems in states in $|\Psi_j\rangle$ with probabilities p_j. The probability of outcome k when a measurement, which is described by P_k, is:

$$k = \text{tr}(\rho P_k) \tag{6.8}$$

where,

$$\rho = \sum_j p_j |\psi_j\rangle\langle\psi_j| \tag{6.9}$$

is the density matrix. ρ completely determines the measurement statistics. We can generalise to an ensemble of non identical states:

$$\{(|\psi_1\rangle, p_1), (|\psi_2\rangle, p_2), ..., (|\psi_k\rangle, p_k)\} \tag{6.10}$$

The set of all probabilities and their associated state vectors $\{p_j, |\psi_j\rangle\}$ is called an ensemble of pure states.

If a measurement is done, with projectors P_k on a system with density matrix ρ the post measurement density matrix ρ_k is:

$$\rho_k = \frac{P_k \rho P_k}{tr(P_k \rho P_k)} . \tag{6.11}$$

A simple example involving the probabilities for a qubit in a known state is next:

Example For a single qubit in the basis $\{|0\rangle, |1\rangle\}$.

$$\rho = \sum_{j=1}^{2} p_j |\psi_j\rangle\langle\psi_j|.$$

For $|\Psi\rangle = 1 \cdot |0\rangle + 0 \cdot |1\rangle$, i.e. measurement probabilities $p_{|0\rangle} = 1$ and $p_{|1\rangle} = 0$:

$$\rho = 1 \cdot |0\rangle\langle 0| + 0 \cdot |1\rangle\langle 1|$$

$$= 1 \cdot \begin{bmatrix} 1 \\ 0 \end{bmatrix} \begin{bmatrix} 1 & 0 \end{bmatrix} + 0 \cdot \begin{bmatrix} 0 \\ 1 \end{bmatrix} \begin{bmatrix} 0 & 1 \end{bmatrix}$$

$$= \begin{bmatrix} 1 & 0 \\ 0 & 0 \end{bmatrix} .$$

For $|\Psi\rangle = 0 \cdot |0\rangle + 1 \cdot |1\rangle$, i.e. measurement probabilities $p_{|0\rangle} = 0$ and $p_{|1\rangle} = 1$:

$$\rho = 0 \cdot |0\rangle\langle 0| + 1 \cdot |1\rangle\langle 1|$$

$$= 0 \cdot \begin{bmatrix} 1 \\ 0 \end{bmatrix} \begin{bmatrix} 1 & 0 \end{bmatrix} + 1 \cdot \begin{bmatrix} 0 \\ 1 \end{bmatrix} \begin{bmatrix} 0 & 1 \end{bmatrix}$$

$$= \begin{bmatrix} 0 & 0 \\ 0 & 1 \end{bmatrix} .$$

Next we'll have a look at a qubit in an unknown state, and the use of a trace over the density matrix given a projector.

Example Given measurement probabilities $p_{|0\rangle} = p$ and $p_{|1\rangle} = 1 - p$.

$$\rho = p|0\rangle\langle 0| + (1-p)|1\rangle\langle 1|$$

$$= p \begin{bmatrix} 1 \\ 0 \end{bmatrix} \begin{bmatrix} 1 & 0 \end{bmatrix} + (1-p) \begin{bmatrix} 0 \\ 1 \end{bmatrix} \begin{bmatrix} 0 & 1 \end{bmatrix}$$

$$= p \begin{bmatrix} 1 & 0 \\ 0 & 0 \end{bmatrix} + 1 - p \begin{bmatrix} 0 & 0 \\ 0 & 1 \end{bmatrix}$$

$$= \begin{bmatrix} p & 0 \\ 0 & 1-p \end{bmatrix}.$$

So, given a density matrix ρ and a projector $P = |0\rangle\langle 0|$ we can extract the final probability from ρ, say we measure in an orthonormal basis, $\{|0\rangle, |1\rangle\}$ then:

$$p_{|0\rangle} = \text{tr}(\rho|0\rangle\langle 0|)$$

$$= \text{tr}\left(\begin{bmatrix} p & 0 \\ 0 & 1-p \end{bmatrix} \begin{bmatrix} 1 & 0 \\ 0 & 0 \end{bmatrix} \right)$$

$$= \text{tr}\left(\begin{bmatrix} p & 0 \\ 0 & 0 \end{bmatrix} \right)$$

$$= p + 0$$

$$= p,$$

$$p_{|1\rangle} = \text{tr}(\rho|1\rangle\langle 1|)$$

$$= \text{tr}\left(\begin{bmatrix} p & 0 \\ 0 & 1-p \end{bmatrix} \begin{bmatrix} 0 & 0 \\ 0 & 1 \end{bmatrix} \right)$$

$$= \text{tr}\left(\begin{bmatrix} 0 & 0 \\ 0 & 1-p \end{bmatrix} \right)$$

$$= 0 + (1-p)$$

$$= 1 - p.$$

How does a the density matrix evolve?

Suppose a unitary transform U is applied to a quantum system: i.e. $U|\Psi\rangle$; what is the new density matrix? To answer this, using the ensemble view, we can say that if the system can be in states $|\Psi_j\rangle$ with probabilities tr_j then, after the evolution occurs, it will be in state $U|\Psi\rangle$ with probabilities p_j.

Initially we have $\rho = \sum_j p_j |\psi_j\rangle\langle\psi_j|$. So, after U is applied, we have:

$$\rho' = \sum_j p_j U |\psi_j\rangle\langle\psi_j| U^\dagger \tag{6.12}$$

$$= U\left(\sum_j p_j |\psi_j\rangle\langle\psi_j|\right) U^\dagger \tag{6.13}$$

$$= U\rho U^\dagger. \tag{6.14}$$

Note that when $|\psi_j\rangle$ goes to $U|\psi_j\rangle$, $\langle\psi_j| = |\psi_j\rangle^\dagger$ goes to $(U|\psi_j\rangle)^\dagger = \langle\psi_j|U^\dagger$.

Example Given measurement probabilities $p_{|0\rangle} = p$ and $p_{|1\rangle} = 1 - p$ then $\rho = \begin{bmatrix} p & 0 \\ 0 & 1-p \end{bmatrix}$.

If X is applied to ρ then:

$$\rho' = X\rho X^\dagger$$

$$= \begin{bmatrix} 1-p & 0 \\ 0 & p \end{bmatrix}.$$

Example For $|\Psi\rangle = \frac{1}{\sqrt{2}}|00\rangle + \frac{1}{\sqrt{2}}|11\rangle$ we have $p_{|00\rangle} = \frac{1}{2}$ and $p_{|11\rangle} = \frac{1}{2}$. This gives us a what is called *completely mixed state* and $\rho = \frac{I}{2}$. So,

$$\rho' = U\frac{I}{2}U^\dagger$$

$$= \frac{I}{2} \text{ because } UU^\dagger = 1.$$

Properties:

$$tr(\rho) = 1. \tag{6.15}$$

ρ is a positive matrix. $\hspace{6cm}$ (6.16)

6.5.4.3 *Subsystem Point of View*

The density matrix can describe any subsystem of a larger quantum system, including mixed subsystems. Subsystems are described by a *reduced density matrix*. If we have two subsystems A and B of a system C where $A \otimes B = C$. The density matrices for the subsystems are ρ^A and ρ^B and the overall density matrix is ρ^C (also referred to as ρ^{AB}). We can define ρ^A and ρ^B as:

$$\rho^A = \text{tr}_B(\rho^C) \text{ and } \rho^B = \text{tr}_A(\rho^C). \tag{6.17}$$

tr_A and tr_B are called *partial traces* over systems A and B respectively. The partial trace is defined as follows:

$$\rho^A = tr_B(|a_1\rangle\langle a_2| \otimes |b_1\rangle\langle b_2|) = |a_1\rangle\langle a_2|tr(|b_1\rangle\langle b_2|) \tag{6.18}$$

$$= \langle b_1|b_2\rangle|a_1\rangle\langle a_2|. \tag{6.19}$$

Previously we mentioned the difference between pure and mixed states. There's a simple test we can do to determine is a state is mixed or pure, which is to run a trace on that state, if we get $tr(\rho^2) < 1$ then the state is mixed ($tr(\rho^2) = 1$ for a pure state). Bell states for example have a pure combined state with $tr((\rho^C)^2) = 1$, but they have mixed substates, i.e. $tr((\rho^A)^2) < 1$ and $tr((\rho^B)^2) < 1$.

6.5.4.4 *Fundamental Point of View*

In terms of the density matrix the four postulates of quantum mechanics are [30]:

1. Instead of using a state vector, we can use the density matrix to describe a quantum system in Hilbert space. If a system is in state ρ_j with a probability of p_j it has a density matrix of $\sum_j p_j \rho_j$.

2. Changes in a quantum system are described by $\rho \to \rho' = U\rho U^\dagger$.
3. Measuring using projectors P_k gives us k with probability $\text{tr}(P_k\rho)$ leaving the system in a post measurement state of $\rho_k = \frac{P_k\rho P_k}{\text{tr}(P_k\rho P_k)}$.
4. A tensor product gives us the state of a composite system. A subsystem's state can be found by doing a partial trace on the remainder of the system (i.e. over the other subsystems making up the system).

6.5.4.5 *Von Neumann Entropy*

The probability distributions in classical Shannon entropy, H, are replaced by a density matrix ρ in Von Neumann entropy, S:

$$S(\rho) = -tr(\rho \log_2 \rho). \tag{6.20}$$

We can also define the entropy in terms of eigenvalues λ_i:

$$S(\rho) = -\sum_i \lambda_i \log_2 \lambda_i \tag{6.21}$$

where λ_i are the eigenvalues of ρ.

If we want to define the uncertainty of a quantum state before measurement we can use entropy. Given a Hilbert space of dimension d then

$$0 \le S(\rho) \le \log_2 d \tag{6.22}$$

with $S(\rho) = 0$ meaning a pure state and $S(\rho) = \log_2 d$ giving us a totally mixed state. For example we could compare two states, by measuring their Von Neumann entropy and determine if one is more entangled than the other. We also use Von Neumann entropy to define a limit for quantum data compression, namely *Schumacher compression*, which is beyond the scope of this tutorial.

Properties:

$$S(\rho_A \otimes \rho_B) = S(\rho_A) + S(\rho_B). \tag{6.23}$$
$$S(\rho_{AB}) \leq S(\rho_A) + S(\rho_B). \tag{6.24}$$
$$S(\rho_{AB}) \geq |S(\rho_A) - S(\rho_B)|. \tag{6.25}$$
$$S(\rho_A) = -\mathrm{tr}(\rho_A \log_2 \rho_A). \tag{6.26}$$
$$S(\rho_B) = -\mathrm{tr}(\rho_B \log_2 \rho_B). \tag{6.27}$$
$$\rho_A = \mathrm{tr}_B(\rho_{AB}). \tag{6.28}$$
$$\rho_B = \mathrm{tr}_A(\rho_{AB}). \tag{6.29}$$
$$\rho_{AB} = \rho_A \otimes \rho_B. \tag{6.30}$$
$$S(A) + S(B) \leq S(AC) + S(BC). \tag{6.31}$$
$$S(ABC) + S(B) \leq S(AB) + S(BC). \tag{6.32}$$

For $S(A) + S(B) \leq S(AC) + S(BC)$ it holds for Shannon entropy since $H(A) \leq H(AC)$ and $H(B) \leq H(BC)$ we get an advantage with Von Neumann entropy with:

$$S(A) > (AC) \tag{6.33}$$

or,

$$S(B) > S(BC). \tag{6.34}$$

It should be noted that quantum mechanics tells us that (6.32) and (6.33) cannot be true simultaneously.

6.6   Noise and Error Correction

6.6.0.6   *Noisy Channels*

Noise is randomisation in a channel. To combat noise we use redundancy, i.e. we send additional information to offset the noise. In the case of a binary channel we use *repetition*, e.g. we can use three 1's to represent a single 1, that way two bits have to be flipped to produce an error. An example of this is that a 011 can equate to 1. If two bit flips is highly improbable then upon receiving 011 we assume (with a high degree of certainty) that a

1 was sent (as 111) and a single bit was flipped. Repetition is inefficient: to make the probability of error occurring lower, longer encodings are required which in turn increases transmission times. Shannon found a better way,

> Given a noisy channel there is a characteristic rate R, such that any information source with entropy less than R can be encoded so as to transmit across the channel with arbitrarily few errors, above R we get errors [16].

So if there is no noise then R matches the channel capacity C.

6.6.0.7 *Classical Error Correction*

We'll consider using *binary symmetric channels* with an error probability p with $p \leq 0.5$ and:

- If 0 is transmitted, 0 is received with probability $1 - p$.
- If 0 is transmitted, 1 is received with probability p.
- If 1 is transmitted, 1 is received with probability $1 - p$.
- If 1 is transmitted, 0 is received with probability p.

We generally use a greater number of bits than the original message to encode the message with *codewords*. We call this a K_x channel coding, with x being the number of bits used to encode the original message.

If we have a code (called a binary block code) K of length n it has an information rate of:

$$R(K) = \frac{k}{n} \tag{6.35}$$

if it has 2^k codewords ($k = 1$ in the example below).

This means, we have an original message of k bits and we use codewords on n bits.

6.6.0.8 *Repetition Codes*

Using *repetition codes* we have a greater chance of success by increasing the number of coding bits for a bit to be transmitted and averaging the result bits. Repetition codes have an information rate of $R(K) = \frac{1}{n}$.

Example A K_3 channel coding could be:

$$0 \to 000,$$

$$1 \to 111.$$

With a channel decoding of:

$$000 \to 0, 001 \to 0, 010 \to 0, 100 \to 0,$$

$$111 \to 1, 110 \to 1, 101 \to 1, 011 \to 1.$$

So our information rate is:

$$R(K_3) = \frac{1}{3} .$$

So, bit flips in 1 out of 3 bits in codeword are fixable, but the information rate is down to $\frac{1}{3}$.

6.6.1 *Quantum Noise*

In practice we cannot make perfect measurements and it's hard to prepare and apply quantum gates to perfect quantum states because real quantum systems are quite noisy.

6.6.2 *Quantum Error Correction*

Quantum error correction codes have been successfully designed, but the area still remains a hot topic. Some scientists still believe that quantum computing may be impossible due to decoherence which is external influences destroying or damaging quantum states. We use quantum error correction codes in place of classical ones. The fact that they are quantum gives us a number of extra problems:

(1) No Cloning.

(2) Continuous errors. Many types of error can occur on a single qubit (not just a bit flip as with a classical circuit). E.g. we might have a change of phase: $\alpha|0\rangle + \beta|1\rangle \to \alpha|0\rangle + e^{i\theta}\beta|1\rangle$.

(3) Measurement destroys quantum information (so if we used a repetition code, how do we apply majority logic to recover the qubits?).

Below are some simple examples of quantum errors.

Example A qubit's relative phase gets flipped:

$$a|0\rangle + b|1\rangle \to a|0\rangle - b|1\rangle.$$

Example A qubit's amplitudes get flipped:

$$a|0\rangle + b|1\rangle \to b|0\rangle + a|1\rangle.$$

Example A qubit's amplitudes and relative phase get flipped:

$$a|0\rangle + b|1\rangle \to b|0\rangle - a|1\rangle.$$

6.6.2.1 *Quantum Repetition Code*

A quantum repetition code is the analogue of a classical repetition code. For classical states (states not in a superposition) this is easy:

$$|0\rangle \to |000\rangle,$$

$$|1\rangle \to |111\rangle.$$

The no cloning theorem won't allow us to make copies of qubits in a super-position, i.e. it prevents us from having:

$$|\Psi\rangle \to |\Psi\rangle|\Psi\rangle|\Psi\rangle$$

which expanded would be:

$$(\alpha|0\rangle + \beta|1\rangle) \otimes (\alpha|0\rangle + \beta|1\rangle) \otimes (\alpha|0\rangle + \beta|1\rangle).$$

So what we do is to encode our superposed state into the following entangled state:

$$|\Psi\rangle = \alpha|0\rangle + \beta|1\rangle \rightarrow \alpha|0\rangle|0\rangle|0\rangle + \beta|1\rangle|1\rangle|1\rangle = |\Psi'\rangle$$

or,

$$\alpha|0\rangle + \beta|1\rangle \rightarrow \alpha|000\rangle + \beta|111\rangle$$

which, expanded is:

$$\alpha|000\rangle + 0|001\rangle + 0|010\rangle + 0|011\rangle + 0|100\rangle + 0|101\rangle + 0|110\rangle + \beta|111\rangle.$$

A simple circuit for this encoding scheme is shown below:

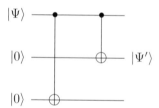

So if we made our input state $\alpha|0\rangle + \beta|1\rangle$ then we would get:

$$\alpha|0\rangle + \beta|1\rangle \qquad |0\rangle \qquad \alpha|000\rangle + \beta|111\rangle$$

$$|0\rangle$$

$$\uparrow |\psi_1\rangle \quad \uparrow |\psi_2\rangle \quad \uparrow |\psi_3\rangle$$

The diagram shows the stages in the evolution of $|\Psi\rangle$ as the CNOT gates are applied; the $|\Psi_i\rangle$ are as follows:

$$|\psi_1\rangle = \alpha|000\rangle + \beta|100\rangle.$$

$$|\psi_2\rangle = \alpha|000\rangle + \beta|101\rangle.$$

$$|\psi_2\rangle = \alpha|000\rangle + \beta|111\rangle.$$

6.6.2.2 *Fixing Errors*

The following circuit detects an amplitude flip (or bit flip) which is the equivalent of $X|\Psi\rangle$ on $|\Psi\rangle$. For example noise can flip qubit three, i.e. $\alpha|000\rangle + \beta|111\rangle \rightarrow I \otimes I \otimes X \rightarrow \alpha|001\rangle + \beta|110\rangle$. We determine that there's been an error by entangling the encoded state with two ancilla qubits and performing measurements on the ancilla qubits. We then adjust our state accordingly based on the results of the measurements as described in the table below. Note that we are assuming that an error has occurred AFTER the encoding and BEFORE we input the state to this circuit.

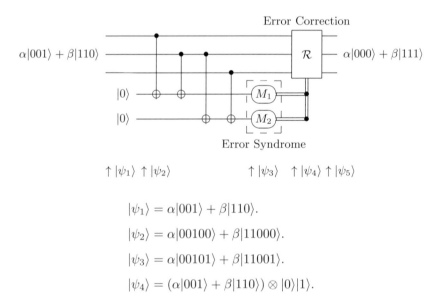

$$|\psi_1\rangle = \alpha|001\rangle + \beta|110\rangle.$$

$$|\psi_2\rangle = \alpha|00100\rangle + \beta|11000\rangle.$$

$$|\psi_3\rangle = \alpha|00101\rangle + \beta|11001\rangle.$$

$$|\psi_4\rangle = (\alpha|001\rangle + \beta|110\rangle) \otimes |0\rangle|1\rangle.$$

The measurements M_1 and M_2 cause a readout of 01 on lines 4 and 5. So now we feed 01 (called the *error syndrome*) into our error correction (or recovery) circuit R which does the following to $\alpha|001\rangle + \beta|110\rangle$:

M_1	M_2	Action
0	0	no action needed, e.g. $|111\rangle \to |111\rangle$
0	1	flip qubit 3, e.g. $|110\rangle \to |111\rangle$
1	0	flip qubit 2, e.g. $|101\rangle \to |111\rangle$
1	1	flip qubit 1, e.g. $|011\rangle \to |111\rangle$

So we apply a qubit flip to line 3 giving:

$$|\psi_5\rangle = \alpha|000\rangle + \beta|111\rangle.$$

This circuit will fix a single bit flip in our three qubit repetition code. All that remains is to decode $|\Psi\rangle$ to return to our original state. We get a problem though if we have a relative phase error, i.e.:

$$\alpha|000\rangle + \beta|111\rangle \to \alpha|000\rangle - \beta|111\rangle$$

which, decoded is:

$$\alpha|0\rangle + \beta|1\rangle \to \alpha|0\rangle - \beta|1\rangle.$$

It turns out we have to change our encoding method to deal with a relative phase flip, which we can do with the following circuit:

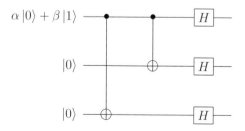

The error correction circuit for a relative phase flip is almost exactly the same as the error correction circuit for an amplitude flip, we just add in some Hadamard gates at the start to deal with the superpositions we generated with the initial encoding (shown immediately above). Again remember that any errors that do happen, happen between the encoding and the circuit we are about to introduce:

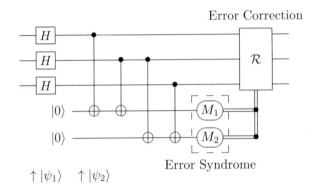

Suppose we have a phase flip on line 2. If our input state is:

$$|\psi_1\rangle = \alpha\,|+ + -\rangle + \beta\,|- - +\rangle$$

then,

$$|\psi_2\rangle = (\alpha\,|001\rangle + \beta\,|110\rangle)\,|00\rangle\,.$$

This is the same as $|\psi_2\rangle$ in the bit flip case. Since the rest of the circuit is the same we get the output of R being:

$$\alpha\,|000\rangle + \beta\,|111\rangle$$

as before.

It should be noted that these errors are defined in terms of the $\{|0\rangle, |1\rangle\}$ basis. If we use the $\{|+\rangle, |-\rangle\}$ basis then the phase flip circuit above fixes a bit flip and vice versa.

In terms of single qubits, a relative phase flip can be fixed with HZH and an amplitude flip with X. But we still have a problem because the circuit above cannot detect an amplitude flip. A third encoding circuit produces a *Shor code* which is a nine qubit code that has enough information for us to be able to apply both types of error correcting circuits, and is our first real *QECC* (Quantum Error Correction Code). The circuit is presented below.

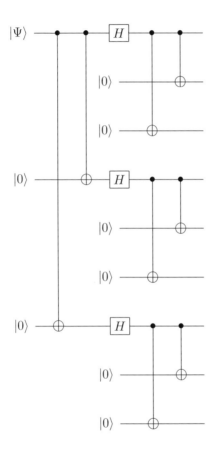

The Shor code is one of many QECCs, another example is the *Steane code*. These both in turn are *CSS codes* (Calderbank, Shor, Steane). CSS codes are part of a more general group of QECCs called *Stabiliser codes*.

6.7   Bell States

As mentioned in chapter 4 the EPR experiment showed that entangled particles seem to communicate certain quantum information instantly over an arbitrary distance upon measurement. Einstein used this as an argument against the notion that particles had no defined state until they were measured (i.e. he argued that they had "hidden variables"). John Bell, 1928–1990 proved in 1964 that there could be no local hidden variables. In the following description of his proof consider the fact that we can measure the spin of a particle in several different directions which can be thought of as measuring in different bases. During this section we'll refer to Einstein's position as **EPR** and the quantum mechanical position as **QM**.

Bell takes a system of two entangled spin particles being measured by two remote observers (observers 1 and 2, say, who observe particles 1 and 2, respectively). He shows that, if the observers measure spin in the same direction (e.g. z) there is no empirical difference between the results predicted by the two positions (**QM** or **EPR**). For the purposes of our example we'll begin by defining the following entangled state:

$$|\Psi_A\rangle = \frac{1}{\sqrt{2}}|0_1 1_2\rangle - \frac{1}{\sqrt{2}}|1_1 0_2\rangle. \qquad (6.36)$$

We've introduced the notation $|0_1 1_2\rangle$ to distinguish between qubits 1 and 2, which will become important later.

If we allow the directions to be different it is possible to get an empirically testable difference. The intuition for this is that if, as **QM** says, what happens at observer 2 is dependent on what happens at observer 1, and the latter is dependent on the direction at 1, then what happens at 2 will involve both directions (we imagine that a measurement at 1, in some direction, causes state collapse so that particle 2 "points" in the antiparallel direction to 1 and is then measured in observer 2's direction; the probability for an outcome, spin up or spin down, is dependent on the angle between the two directions — see below). Whereas, in the **EPR** view, what happens at observer 2 cannot in any way involve the direction at observer 1: what happens at observer 2 will, at most, be determined by the value particle 2 carries away at emission and the direction at observer 2.

6.7.1 *Same Measurement Direction*

To see that same direction measurements cannot lead to a testable difference first imagine that there are two separated observers, Alice and Bob, who run many trials of the **EPR** experiment with spin particles. Let each measure the spin of the particle that flies towards them. If they both measure in the same direction, say z, then the results might be represented as follows:

Alice	*Bob*	**Frequency**
z	z	
1	0	50%
0	1	50%

We soon realise that we can't use this case to discriminate between the two positions.

QM position. This explains the result by saying that, upon Alice's measurement of 1, the state vector superposition $|\Psi_A\rangle$ collapses into one or other of the two terms.

$$\text{either } |0_1 1_2\rangle \text{ or } |1_1 0_2\rangle$$

with 50% probability for Alice measuring 0 or 1. But then we no longer have a superposition (Alice's measurement has "produced" a 0 on particle 1 and a 1 on particle 2, or vice versa) and Bob's measurement outcome on 2 is completely determined (relative to the outcome on 1) and will be the opposite of Alice's outcome.

EPR position. This says the particles actually have well defined values at the moment they are emitted, and these are (anti-)correlated and randomly distributed between 01 and 10 over repeated trials. When Alice measures 0 Bob measures 1 not because Alice's measurement has "produced" anything; but simply because the 1 is the value the particle had all along (which the measurement reveals). The randomness here is due to initial conditions (emission); not to collapse.

Either way we have the same prediction.

6.7.2   *Different Measurement Directions*

Bell's innovation was to realise that a difference will occur if we bring in different measurement directions. The intuition for this might go as follows.

Suppose there are two directions $a(=z)$ and b. Start with $|\Psi_A\rangle$ and allow Alice to measure particle 1 in the a direction. This causes a collapse, as before, and we have:

$$\text{either } |0_1 1_2\rangle \text{ or } |1_1 0_2\rangle.$$

Alice's measurement has "produced" a 1 (spin down) in the a direction on particle 2 (to take the first case). Bob now measures particle 2 in the b direction. To work out what happens according to QM we have to express $|1_2\rangle$ in terms of a b basis. If b is at angle θ to a this turns out to be:

$$|1_2\rangle = \sin\left(\frac{\theta}{2}\right)|+_2\rangle + \cos\left(\frac{\theta}{2}\right)|-_2\rangle \qquad (6.37)$$

where $|+_2\rangle$ means spin up along the b axis. This looks right. If $\theta = 90°$ (the case when $b = x$) we get $\frac{1}{\sqrt{2}}$, as before; when $\theta = 0$ (the case when $b = z$) we get $|-_2\rangle$ $(= |1\rangle$ in that case). Likewise, for $|0_2\rangle$:

$$|0_2\rangle = \cos\left(\frac{\theta}{2}\right)|+_2\rangle - \sin\left(\frac{\theta}{2}\right)|-_2\rangle. \qquad (6.38)$$

The significant point is that, since θ is the angle between a and b, the direction of Alice's measurement is entering into (influencing) what happens for Bob.

But this can't happen in a realist local view as, espoused by **EPR**, because:

(1) The spin values of the particles are determined at the moment of emission, and are not "produced" at some later moment as a result of an act of measurement (realism);
(2) by construction, the events at Bob's end are beyond the reach of the events at Alice's end (locality).

The rat is cornered: by bringing different directions of measurement into it, we should get a detectable difference. So reasons Bell.

6.7.3 *Bell's Inequality*

It turns out we need three directions. So allow Alice and Bob to measure in three possible directions labeled a, b, and c; we will actually take the case where these directions are all in the plane, perpendicular to line of flight, and at $120°$ to one another. When Alice and Bob measure in the same direction, say a and a, they will observe correlation: 1 with 0, or 0 with 1, as above. However, if they measure in different directions, say a and b, there is no requirement for correlation, and Alice and Bob might both observe 1, for example (as in row 3 below).

6.7.3.1 *QM View*

If we put $\theta = 120$ into 6.38 we have:

$$|1_2\rangle = \frac{\sqrt{3}}{2}|+_2\rangle + \frac{1}{2}|-_2\rangle.$$

This tells us that Alice having measured 0 in the a direction. Bob now measures $+$ in the b (i.e. spin up or 0) direction with probability $\frac{3}{4}$ and $-$ in the b direction (i.e. spin down or 1) with probability $\frac{1}{4}$. The net result is that we have:

Alice	*Bob*	**Probability**
a	b	
0	0	$\frac{3}{4}$
0	1	$\frac{1}{4}$

Equivalent results will be obtained in other pairs of different directions (e.g. bc, i.e. Alice chooses b, Bob c) since the angles between such directions are always the same. In general we have:

> If the directions are different the probability for outcomes to be the same S (e.g. 00) is $\frac{3}{4}$; to be different D (e.g. 01) is $\frac{1}{4}$. On the other hand, if the directions are the same the probability for different outcomes (D) is 1.

Suppose we now run 900 trials. Randomly change the directions of measurement, so that each combination occurs 100 times. The results must go as:

	aa	ab	ac	ba	bb	bc	ca	cb	cc
Outcome S	0	75	75	75	0	75	75	75	0
Outcome D	100	25	25	25	100	25	25	25	100

The net result is that S occurs 450 times and D 450 times. We conclude:

The probability of different outcomes is $\frac{1}{2}$.

6.7.3.2 *EPR View*

We assume each particle carries a spin value relative to each possible direction (as in classical physics), except that (to fit in with what is observed) the measured value will always be "spin up" (0) or "spin down" (1), and not some intermediate value (so the measured values cannot be regarded as the projection of a spin vector in 3D space onto the direction of the magnetic field, as in classical physics). To try to account for this we say that, at the moment of emission, each particle leaves with an "instruction" (the particle's property, or hidden variable, or key, or component) — one for each possible direction (a, b, c) in which it can be measured — which determines the measured outcome in that direction (in the conventional, pre-quantum, sense that the measurement just discovers or reveals what is already there). Since the measurement in the a direction can discover two possible values, 0 or 1, the corresponding property, which we call the A component, must have two values: either A=0, or A=1, which determines the measurement outcome (as noted above, we cannot allow randomness in the act of measurement — this would destroy the aa correlation, etc. The only possible randomness that can occur is in the assignment of initial component values at the moment of emission, see below). Likewise, we have B=0 or B=1, and C=0 or C=1. (It might seem we need an infinity of these properties to cater for the infinity of possible measurement directions. This doesn't constitute a problem in classical physics where we have spin behaving like a magnetic needle: the infinity of possible observed components of the magnetic moment, dependent on measurement direction, is no more mysterious than the infinity of shadow lengths projected by a stick dependent on the sun's direction. But note that it would seem strange to account for the infinity of shadow length's by saying the stick had an infinity of components — or "instructions" — specifying what shadow should appear, depending on the angle of the sun!).

The A,B,C values are set for each particle at the time of emission and carried off by them. They can be randomly assigned except that, to obey conservation, if A=0 for particle 1 then A=1 for particle 2 etc. But we can have A=0 for particle 1 and B=0 for particle 2 since, empirically, the spins are not correlated if we measure in different directions (in this case if the measurements were in the a and b directions the result would be 00).

Suppose particle 1 has the assignment A=0, B=1, C=1; i.e. 011. Interestingly, we immediately know what the components for particle 2 must be: A=1, B=0, C=0; i.e. 100. This is necessary if the two particles are going to give anti-correlated results for the case where we do same direction measurements in each direction (a, b, c).

We can then lay out the possible assignments of component values, as follows:

Row	Alice			Bob		
	a	b	c	a	b	c
1.	1	1	1	0	0	0
2.	0	1	1	1	0	0
3.	1	0	1	0	1	0
4.	0	0	1	1	1	0
5.	1	1	0	0	0	1
6.	0	1	0	1	0	1
7.	1	0	0	0	1	1
8.	0	0	0	1	1	1

The table respects the rule that measurements in the same direction must be correlated (this is how the entries are generated — by allowing Alice's *abc* values to run over the 8 binary possibilities; for each case the correlation rule immediately determines the corresponding Bob entries — i.e. we have no choice). But it allows for all possibilities in different directions: e.g. we can have Alice finding 1 in the a direction going with Bob finding 1 in this direction (rows 3 and 7) or 0 (rows 1 and 5).

We are now closing in. Suppose we consider the above 900 trials from the realist perspective. There are only eight possible assignments of properties (nature's choice). And there are nine possible combinations of measurement directions (the experimenters' choice). We are assuming that the particles fly off with their assignments and the measurements simply reveal these (no disturbance to worry about, as agreed). So in each case we can tell the outcome. For example, if Nature chooses row 7 (above), i.e. 100 and the experimenters choose ab then the outcome is completely determined as S, as follows. Alice gets a 100 particle, measures it in the a direction, and gets 1, Bob gets a 011 particle, measures it in the b direction, and gets 1. Hence the overall outcome is 11, i.e. S.

We can now complete the table of outcomes:

Alice's particle	*aa*	*ab*	*ac*	*ba*	*bb*	*bc*	*ca*	*cb*	*cc*
111	D	D	D	D	D	D	D	D	D
011	D	S	S	S	D	D	S	D	D
101	D	S	D	S	D	S	D	S	D
001	D	D	S	D	D	S	S	S	D
110	D	D	S	D	D	S	S	S	D
010	D	S	D	S	D	S	D	S	D
100	D	S	S	S	D	D	S	D	D
000	D	D	D	D	D	D	D	D	D

What is the probability of different outcomes D? If Alice receives a particle with equal components (e.g. 111) this probability is always 1. If Alice receives a particle with unequal components (e.g. 011) the probability is always $\frac{5}{9}$ (look along the row for 011: D occurs 5 times out of 9. We don't know the statistics of the source: how often particles of the 8 various kinds are emitted. But we can conclude that the overall probability for D is:

$$p_{\text{equal}} + \frac{5}{9} p_{\text{unequal}}.$$

Since both p's must be positive we conclude: the probability for different outcomes is greater than $\frac{5}{9}$.

This is an example of a Bell inequality.

6.7.3.3 *Contradiction!*

If we compare the two conclusions we have a contradiction. **EPR** says the probability should be $\frac{1}{2}$; experiment shows it should be greater than $\frac{5}{9}$ so **QM** wins!

6.8 Cryptology

Now a proven technology, quantum cryptography provides total security between two communicating parties. Also, the unique properties of quantum computers promise the ability to break classical encryption schemes like Data Encryption Standard (DES) and RSA. The field of quantum cryptology has two important sub-fields, they are:

(1) **Cryptography** — The use of secure codes.
(2) **Cryptanalysis** — Code breaking.

We'll concentrate on cryptography in this chapter. Later, in chapter 7 we look at Shor's algorithm, which can be used to break RSA encryption.

6.8.1 *Classical Cryptography*

Secret codes have a long history — dating back to ancient times. One famous ancient code is the *Caesar cipher* which simply shifts each letter by three.

$$A \rightarrow D, B \rightarrow E, X \rightarrow A. \tag{6.39}$$

This is not very secure as it's easy to guess and decrypt. Modern codes use a key. A simple form of key is incorporated into a *code wheel*, an example of which follows.

Example An example of a code wheel with the key: ABCDEF is described below:

Key	A	B	C	D	E	F	A	B	C	D	E	F
Shift By	1	2	3	4	5	6	1	2	3	4	5	6
Message	Q	U	A	N	T	U	M	C	O	D	E	S
Encoding	R	W	D	R	Y	A	N	E	R	H	J	Y

With time secret keys and their respective codes got more complicated, culminating in modern codes, e.g. codes called DES and IDEA were implemented with typically 64 and 128 bit secret keys.

There is another kind of key, a public key. *Public key encryption* (e.g. RSA 1976 Diffie Hellman) uses a combination of a public and a secret key. In a public key encryption scheme a public key encrypts a message but cannot decrypt it. This is done by a secret decryption key known only to the owner. Surmising the decryption key from public (encryption) key requires solving hard (time complexity wise) problems.

Some codes are stronger than others, for example the DES and IDEA secret key codes are stronger than Caesar cipher codes. The strongest code is the *one-time PAD*. A one-time PAD is a key that is large as the message to be sent (for example a random binary file) and the sender and receiver both have a copy of this key. The PAD is used only once to encrypt and decrypt the message. The problem with one-time PADs is that the key needs to be transmitted every time and eavesdroppers could be listening in on the key. Quantum key distribution resolves this issue by providing perfectly secure key distribution.

6.8.2 *Quantum Cryptography*

Modern classical public key cryptographic systems use a *trap door function* with security based on mathematical assumptions, notably that it is difficult to factor large integers. This assumption is now at risk from Shor's algorithm.

The main advantage of quantum cryptography is that it gives us perfectly secure data transfer. The first successful quantum cryptographic device was tested in 1989 by C.H. Bennet and G. Brassard [11]. The device could translate a secret key over 30 centimeters using polarised light, calcite crystal(s), and other electro-optical devices. This form of cryptography does not rely on a trap door function for encryption, but on quantum effects like the no-cloning theorem.

A simple example of the no-cloning theorem's ability to secure data is described below. After that we'll look at why it's impossible to listen in on a quantum channel, and finally we'll examine a method for quantum key distribution.

6.8.2.1 *Quantum Money*

Stephen Wiesner wrote a paper in 1970 (unpublished until 1983) in which he described uncounterfeitable quantum bank notes [8].

Each bank note contains a unique serial number and a sequence of randomly polarised photons ($90°$, $180°$, $45°$, and $135°$). The basis of polarisation is kept secret (either diagonal or rectilinear). The bank can verify the validity of a note by matching a serial number with known (only known by the bank) polarisations. This enables the bank to verify the polarisations without *disturbing* the quantum system (see the example below). Finally, the counterfeiter can't counterfeit the note due to the no cloning theorem. Remember, the no cloning theorem says that given an unknown quantum state it is impossible to clone that state exactly without disturbing the state.

Example Let's say the bank receives a bank note with serial number 1573462. The bank checks its "quantum bank note archive" to find the polarisations for bank note 1573462. The banks' records give the following:

$$1573462 = \{H, H, 135°, V, 45°\}.$$

$\{H, H, 135°, V, 45°\}$ is a record of the note polarisations, but our copy protection system does not distinguish between H and V or $135°$ and $45°$. Because the photons are not in superpositions relative to the basis we measure in (as specified by the serial numbers) the bank doesn't disturb the states upon measurement. The bank only determines the basis of each photon, which is $\{|0\rangle, |1\rangle\}$, (rectilinear) or $\{|+\rangle, |1\rangle\}$, (diagonal). It should be noted that states are always superpositions if we don't say what the basis is.

Because the counterfeiter does not know the basis of each polarised photon on the quantum bank note, he must measure using a random basis. He could, theoretically measure each one and recreate it if he knows the basis by which to measure each one. The chances of the counterfeiter randomly measuring in the correct basis decrease by a $\frac{1}{2}$ for each successive polarised photon on the note. So the more polarised photons the bank note has the harder it is to counterfeit.

6.8.2.2 *Quantum Packet Sniffing*

Quantum *packet sniffing* is impossible — so the sender and receiver can be sure that no-one is listening in on their messages (eavesdropping). This phenomena is due to the following properties of quantum mechanics:

(1) Quantum uncertainty, for example given a photon polarised in an un-

known state (it might be horizontally (180°), vertically (90°), at 45°, or at 135°) we can't tell with certainty which polarisation the photon has without measuring it.

(2) The no cloning theorem.

(3) Disturbance (information gain). If a measurement is done (distinguishing between two nonorthogonal states) the signal is forever disturbed.

(4) Measurements are irreversible; any nonorthogonal state will collapse randomly into a resultant state — losing the pre-measurement amplitudes.

6.8.2.3   *Quantum Key Distribution*

We can ensure secure communications by using one-time pads in conjunction with *quantum key distribution*. The main drawback for classical one-time pads is the distribution of encryption/decrytion keys. This is not a problem for quantum cryptography as we can transfer key data in a totally secure fashion.

Quantum key distribution (QKD) is a means of distributing keys from one party to another, and detecting eavesdropping (it is inspired by the quantum money example). An unusually high error rate is a good indicator of eavesdropping. Even with a high error rate the eavesdropper cannot learn any useful information. It should be noted that this method does not prevent Denial of Service (DOS) attacks.

An example follows on the next page.

Example   Alice and Bob want to use a key based encryption method. They communicate on two channels. One channel is used to transmit the encrypted message and another is used to transmit the encryption key. But they need a way to avoid Eve eavesdropping on their conversation on the key channel [19].

The following steps allow a totally secure (in terms of eavesdropping) transfer of the key (later we need to encrypt the message and send it via a classical channel).

(1) Alice randomly polarises photons at 45°, 90° ,135°, and 180° — these are sent to Bob over a quantum channel (which is secure due to quantum effects, i.e. no eavesdropping).

(2) Bob does measurements on the photons. He randomly uses either a rectilinear (180°/90°) or diagonal polarising filter (45°/135°), and records which one was used for each bit. Also for each bit he records the measurement result as a 1 for 135° or 180° or 0 for 45° or 90° polarisations.

(3) On a normal (insecure channel) Alice tells Bob which bits Bob used the right polarisation on (which he also communicates over the classical channel).

(4) Alice and Bob check for quantum errors to gauge whether or not an eavesdropper (Eve) was listening.

6.8.2.4   *Comments*

The following can be said about the key exchange:

- Eve can't clone the bit and send it on; her measurements put Bob's bits in a new random state. Alice and Bob decide on an acceptable error rate, and if the key exchange's error rate is higher than that (which means Eve may have been listening) then they resend the message.
- Eve could do an DOS attack by constantly measuring on the key channel.
- Eve could shave off some of the light if each bit is represented by more than one photon, so each bit transferred must only be represented by one qubit.
- Eve could listen to a small number of bits and hope to be unnoticed — Alice and Bob can prevent this by shrinking down their key.
- In 1984 Bennet and Brassard developed *BB84*, a method of *mutual key generation*, the first QKD system which is based on concepts similar to the above.

6.9 Alternative Models of Computation

There are other candidates for physical phenomena to enhance computational power. Some disciplines cannot currently be described by quantum mechanics, like relativity and complexity theory (or chaos theory).

Question *Is there a universal computation model? Or will there always be some phenomena which is not well understood that has the potential to be exploited for the purposes of computation?*

Chapter 7

Quantum Algorithms

7.0.1   *Introduction*

Quantum algorithms are ways of combining unitary operations in a quantum system to achieve some computational goal. Over the past twenty five years a number of algorithms have been developed to harness the unique properties offered by quantum computers. These algorithms have been designed to give special advantages over their classical counterparts.

Shor's algorithm (1995) gives the factorisation of arbitrarily large numbers a time complexity class of $O((\log N)^3)$, while the classical equivalent is roughly exponential. This is extremely important for cryptography. For example, RSA relies on factorisation being *intractable*, which means that no polynomial solution exists that covers all instances of the problem. Another example is Grover's database search algorithm which provides a quadratic speedup when searching a list of N items. This takes the classical value of a linear search from $O(N)$ time to $O(N^{\frac{1}{2}})$ time. Most known quantum algorithms have similarities to, or partially borrow from, these two algorithms.

Some properties of Shor and Grover type algorithms are:

- Shor type algorithms use the *quantum Fourier transform*. These include: factoring, the hidden subgroup problem, discrete logarithms, and order finding (all variations on a theme).
- Grover type search algorithms can be used for applications like fast database searching and statistical analysis. Simon's algorithm is possibly the first in this group of algorithms.

There are also hybrid algorithms like quantum counting that combine elements from both, and more esoteric algorithms like quantum simulators.

In this chapter we'll look at Deutsch's algorithm, the Deutsch–Josza algorithm, Shor's algorithm, and Grover's algorithm.

As with chapters 5 and 6 individual references to QCQI have been dropped.

7.1 Deutsch's Algorithm

Deutsch's algorithm is a simple example of quantum parallelism. The problem it solves is not an important one, but its simple nature makes it good for demonstrating the properties of quantum superposition.

7.1.1 The Problem Defined

We have a function $f(x)$ where $f(x) : \{0, 1\} \to \{0, 1\}$ with a one bit domain. This means that $f(x)$ takes a bit (either a 0 or a 1) as an argument and returns a bit (again, either a 0 or a 1). Therefore both the input and output of this function can be represented by a bit, or a qubit.

We want to test if this function is one to one (balanced), where one to one means:

$$f(1) = 1 \text{ and } f(0) = 0 \text{ or } f(0) = 1 \text{ and } f(1) = 0.$$

The other alternative is that $f(x)$ is not one to one (i.e. it is constant), in which case we would get:

$$f(1) = 0 \text{ and } f(0) = 0 \text{ or } f(0) = 1 \text{ and } f(1) = 1.$$

The circuit for determining some property of a function is called an *oracle*. It's important to note that the thing that quantum computers do well is to test *global* properties of functions, not the results of those functions given particular inputs. To do this efficiently we need to look at many values simultaneously.

7.1.2 The Classical Solution

The solution is as follows:

The function is one to one iff $f(0) \oplus f(1) = 1$ (where $\oplus$ is equivalent to XOR).

This is made up of three operations, involving two function evaluations:

(1) $x = f(0)$,
(2) $y = f(1)$,
(3) $z = x \oplus y$.

Can we do better with a quantum computer?

7.1.3 The Quantum Solution

The quantum circuit below performs the first two steps in the classical solution in one operation, via superposition.

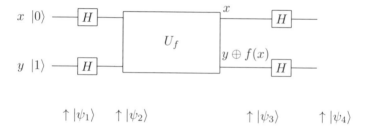

Below are the steps $|\psi_1\rangle$ to $|\psi_4\rangle$ we need to solve the problem quantum mechanically.

$|\psi_1\rangle$ The qubits x and y have been set to the following (we call x the query register and y the answer register):

$$x = |0\rangle,$$
$$y = |1\rangle.$$

$|\psi_2\rangle$ Apply H gates to the input registers so our state vector is now:

$$|\Psi\rangle = \left[\frac{|0\rangle + |1\rangle}{\sqrt{2}} \right] \left[\frac{|0\rangle - |1\rangle}{\sqrt{2}} \right].$$

$|\psi_3\rangle$ U_f acts on qubits x and y, specifically we use $|\Psi\rangle \rightarrow U_f \rightarrow$ $|x\rangle |y \oplus f(x)\rangle$. After some algebraic manipulation the result ends up in x, and y seems unchanged (see the example below).

For $f(0) \neq f(1)$ (balanced) we get:

$$|\Psi\rangle = \pm \left[\frac{|0\rangle - |1\rangle}{\sqrt{2}} \right] \left[\frac{|0\rangle - |1\rangle}{\sqrt{2}} \right]$$

and for $f(0) = f(1)$ (constant):

$$|\Psi\rangle = \pm \left[\frac{|0\rangle + |1\rangle}{\sqrt{2}} \right] \left[\frac{|0\rangle - |1\rangle}{\sqrt{2}} \right].$$

Note: the $\pm$ at the start of the state is just a result of internal calculations, it does not matter whether it is positive or negative for the result of this circuit (global phase factors have no significance).

$|\psi_4\rangle$ x is sent through the H gate again. The H gate, as well as turning a 1 or a 0 into a superposition, will take a superposition back to a 1 or a 0 (depending on the sign):

$$\left[\frac{|0\rangle + |1\rangle}{\sqrt{2}} \right] \rightarrow H \rightarrow |0\rangle,$$

$$\left[\frac{|0\rangle - |1\rangle}{\sqrt{2}} \right] \rightarrow H \rightarrow |1\rangle.$$

So at step 4, for $f(0) \neq f(1)$ we get:

$$|\Psi\rangle = \pm|1\rangle \left[\frac{|0\rangle - |1\rangle}{\sqrt{2}} \right]$$

and for $f(0) = f(1)$:

$$|\Psi\rangle = \pm|0\rangle \left[\frac{|0\rangle - |1\rangle}{\sqrt{2}} \right].$$

At this point our state is, combining both cases:

$$|\Psi\rangle = \pm|f(0) \oplus f(1)\rangle \left[\frac{|0\rangle - |1\rangle}{\sqrt{2}}\right].$$

Up until now y has been useful to us, but now we don't care about it as x holds the result. We can just do a partial measurement on x and discard y as garbage. If $x = 0$ then the function is constant, and if $x = 1$ the function is balanced.

Example   There are only four possible combinations for $f(x)$. We'll look at two here. Given our state at $|\psi_2\rangle$ is $\frac{1}{\sqrt{2}}[|0\rangle + |1\rangle]\frac{1}{\sqrt{2}}[|0\rangle - |1\rangle]$ our state at $|\psi_3\rangle$ will look like the following:

$$|\Psi\rangle = \left[\frac{|0\rangle + |1\rangle}{\sqrt{2}}\right]\left[\frac{|0 \oplus f(x)\rangle - |1 \oplus f(x)\rangle}{\sqrt{2}}\right]$$
$$= \frac{1}{2}(|0\rangle\,|0 \oplus f(0)\rangle - |0\rangle\,|1 \oplus f(0)\rangle + |1\rangle\,|0 \oplus f(1)\rangle - |1\rangle\,|1 \oplus f(1)\rangle).$$

Keeping in mind that $0 \oplus 0 = 0, 1 \oplus 0 = 1, 1 \oplus 1 = 0$, and $0 \oplus 1 = 1$ we'll start with the constant function $f(0) = 0$ and $f(1) = 0$:

$$|\Psi\rangle = \frac{1}{2}(|0\rangle\,|0 \oplus 0\rangle - |0\rangle\,|1 \oplus 0\rangle + |1\rangle\,|0 \oplus 0\rangle - |1\rangle\,|1 \oplus 0\rangle)$$
$$= \frac{1}{2}(|00\rangle - |01\rangle + |10\rangle - |11\rangle)$$
$$= \left[\frac{|0\rangle + |1\rangle}{\sqrt{2}}\right]\left[\frac{|0\rangle - |1\rangle}{\sqrt{2}}\right].$$

Now for a balanced function $f(0) = 1$ and $f(1) = 0$:

$$|\Psi\rangle = \frac{1}{2}(|0\rangle\,|0 \oplus 1\rangle - |0\rangle\,|1 \oplus 1\rangle + |1\rangle\,|0 \oplus 0\rangle - |1\rangle\,|1 \oplus 0\rangle)$$
$$= \frac{1}{2}(|01\rangle - |00\rangle + |10\rangle - |11\rangle)$$
$$= \frac{1}{2}(-|00\rangle + |01\rangle + |10\rangle - |11\rangle)$$
$$= (-1)\frac{1}{2}(|00\rangle - |01\rangle - |10\rangle + |11\rangle)$$
$$= -1\left[\frac{|0\rangle - |1\rangle}{\sqrt{2}}\right]\left[\frac{|0\rangle - |1\rangle}{\sqrt{2}}\right].$$

Notice that (-1) has been moved outside of the combined state; it is a global phase factor and we can ignore it. At $|\psi_4\rangle$ we put our entire state through $H \otimes H$ and get $|01\rangle$ for the constant function and $|11\rangle$ for the balanced one.

7.1.4 *Physical Implementations*

In theory a simple quantum computer implementing Deutsch's algorithm can be made out of a cardboard box, three mirrors, and two pairs of sunglasses [36]. The sunglasses (in a certain configuration) polarise light initially into a nonorthogonal state (a superposition) then back again. The mirrors within the box reflect the light in a certain way depending on their configuration (at different angles depending on the type of function you want to simulate). A torch is then shone into the box. If light comes out the function is balanced, if not then it is constant. Now, if we use a special optical device to send just one photon into the box and we have a sensitive photo detector at the other end then we have an architecture that is a theoretically more efficient oracle than any classical computer.

Deutsch's algorithm can of course be executed on any quantum computer architecture, and has been successfully implemented. For example, in 2001 the algorithm was run on an NMR computer (see chapter 8) [15]. Due to the relatively small number of qubits that can be currently made to work together many of the other quantum algorithms have not been satisfactorily tested.

7.2 The Deutsch–Josza Algorithm

The Deutsch–Josza algorithm is an extension of Deutsch's algorithm which can evaluate more than one qubit in one operation. We can extend it to evaluate any number of qubits simultaneously by using an n-qubit query register for x instead of a single qubit.

7.2.1 *The Problem Defined*

The problem we are trying solve is slightly different to the one presented in Deutsch's algorithm, that is: is $f(x)$ the same (constant) for all inputs? Or is $f(x)$ equal to 1 for half the input values and equal to 0 for the other half (which means it is balanced)?

7.2.2   *The Quantum Solution*

The circuit looks like this:

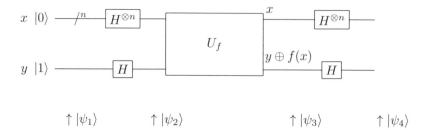

$$\uparrow |\psi_1\rangle \qquad\qquad \uparrow |\psi_2\rangle \qquad\qquad\qquad \uparrow |\psi_3\rangle \qquad\qquad \uparrow |\psi_4\rangle$$

The input (x) and output (y) registers have an H gate for each qubit in the register. This is denoted by $H^{\otimes n}$ and the $/^n$ notation just means n wires for n qubits.

Here are the steps $|\psi_1\rangle$ to $|\psi_4\rangle$ we need to solve the problem quantum mechanically.

$|\psi_1\rangle$ The registers x and y have been set to the following:

$$x = |0\rangle^{\otimes n},$$
$$y = |1\rangle.$$

$|\psi_2\rangle$ Apply H gates to both the x and y registers so our state vector is now:

$$|\Psi\rangle = \frac{1}{\sqrt{2^n}} \sum_{x=0}^{2^n-1} |x\rangle \left[\frac{|0\rangle - |1\rangle}{\sqrt{2}} \right].$$

$|\psi_3\rangle$ U_f acts on registers x and y (remember y is just a single qubit because the result of evaluating f is, by definition, just a 0 or 1). This time we use $|\Psi\rangle = U_f |\Psi_2\rangle |0\rangle = |x_1, x_2, \ldots, x_n\rangle |y \oplus f(x)\rangle$ to evaluate $f(x)$.

$|\psi_4\rangle$ x is sent through $H^{\oplus n}$ and y through an H gate also. This leaves us with a set of output qubits and a 1 in the answer register, i.e.

$$|\Psi\rangle = |x_1, x_2, \ldots, x_n\rangle |1\rangle.$$

Now the rule is simple: if any of the qubits in the query (x) register are $|1\rangle$ then the function is balanced, otherwise it is constant.

Example Here's an example with a two qubit query register for the balanced function f with $f(00) = 0$, $f(01) = 0$, $f(10) = 1$, and $f(11) = 1$. So at state $|\psi_1\rangle$ we have:

$$|\Psi\rangle = |001\rangle .$$

Then at state $|\psi_2\rangle$, after the H gates we get:

$$|\Psi\rangle = \left[\frac{|0\rangle + |1\rangle}{\sqrt{2}}\right]\left[\frac{|0\rangle + |1\rangle}{\sqrt{2}}\right]\left[\frac{|0\rangle - |1\rangle}{\sqrt{2}}\right].$$

We are now ready to examine our state at $|\psi_3\rangle$:

$$\begin{aligned}
|\Psi\rangle =&\left[\frac{|0\rangle + |1\rangle}{\sqrt{2}}\right]\left[\frac{|0\rangle + |1\rangle}{\sqrt{2}}\right]\left[\frac{|0 \oplus f(x)\rangle - |1 \oplus f(x)\rangle}{\sqrt{2}}\right]\\
=&\left[\frac{|00\rangle + |01\rangle + |10\rangle + |11\rangle}{\sqrt{4}}\right]\left[\frac{|0 \oplus f(x)\rangle - |1 \oplus f(x)\rangle}{\sqrt{2}}\right]\\
=&\frac{1}{\sqrt{8}}(|00\rangle |0 \oplus f(00)\rangle - |00\rangle |1 \oplus f(00)\rangle + |01\rangle |0 \oplus f(01)\rangle\\
&- |01\rangle |1 \oplus f(01)\rangle + |10\rangle |0 \oplus f(10)\rangle - |10\rangle |1 \oplus f(10)\rangle\\
&+ |11\rangle |0 \oplus f(11)\rangle - |11\rangle |1 \oplus f(11)\rangle)\\
=&\frac{1}{\sqrt{8}}(|000\rangle - |001\rangle + |010\rangle - |011\rangle + |101\rangle - |100\rangle + |111\rangle - |110\rangle)\\
=&\frac{1}{\sqrt{8}}(|000\rangle - |001\rangle + |010\rangle - |011\rangle - |100\rangle + |101\rangle - |110\rangle + |111\rangle)\\
=&\left[\frac{|00\rangle + |01\rangle - |10\rangle - |11\rangle}{\sqrt{4}}\right]\left[\frac{|0\rangle - |1\rangle}{\sqrt{2}}\right]\\
=&\left[\frac{|0\rangle - |1\rangle}{\sqrt{2}}\right]\left[\frac{|0\rangle + |1\rangle}{\sqrt{2}}\right]\left[\frac{|0\rangle - |1\rangle}{\sqrt{2}}\right].
\end{aligned}$$

At $|\psi_4\rangle$ we put our entire state through $H \otimes H \otimes H$ and get $|101\rangle$ meaning that our function is balanced.

7.3   Shor's Algorithm

7.3.1   *The Quantum Fourier Transform*

The quantum analogue of the discrete Fourier transform (see chapter 4) is the Quantum Fourier Transform (QFT). The DFT takes a series of N complex numbers, $X_0, X_1, ..., X_{N-1}$ and produces a series complex numbers $Y_0, Y_1, ..., Y_{N-1}$. Similarly, the QFT takes a state vector:

$$|\Psi\rangle = \alpha_0|0\rangle + \alpha_1|1\rangle + ... + \alpha_{N-1}|N-1\rangle \tag{7.1}$$

and performs a DFT on the amplitudes of $|\Psi\rangle$ giving us:

$$|\Psi\rangle = \beta_0|0\rangle + \beta_1|1\rangle + ... + \beta_{N-1}|N-1\rangle. \tag{7.2}$$

The main advantage of the QFT is the fact that it can do a DFT on a superposition of states. This can be done on a superposition like the following:

$$\frac{1}{\sqrt{4}}(|00\rangle + |01\rangle - |10\rangle - |11\rangle)$$

or a state where the probability amplitudes are of different values, e.g. the following state which has probability amplitudes of 0 for the majority of its basis states:

$$\frac{1}{\sqrt{3}}(|001\rangle + |011\rangle - |111\rangle).$$

It'll helpful over the next few sections to use integers when we are describing states with a large number of qubits, so the previous state could have been written as:

$$\frac{1}{\sqrt{3}}(|1\rangle + |3\rangle - |7\rangle).$$

The QFT is a unitary operator, and is reversible. In fact we use the *inverse quantum fourier transform* ($QFT^\dagger$) for Shor's algorithm. The QFT

is defined as follows:

Given a state vector $|\Psi\rangle$:

$$|\Psi\rangle = \sum_{x=0}^{2^n-1} \alpha_x |x\rangle \tag{7.3}$$

where n is the number of qubits, $QFT |\Psi\rangle$ is defined as:

$$|\Psi'\rangle = QFT |\Psi\rangle = \sum_{x=0}^{2^n-1} \sum_{y=0}^{2^n-1} \frac{\alpha_x e^{2\pi ixy/2^n}}{\sqrt{2^n}} |y\rangle. \tag{7.4}$$

We can also represent the QFT as a matrix:

$$\frac{1}{\sqrt{2^n}} \begin{bmatrix} 1 & 1 & 1 & \cdots & 1 \\ 1 & \omega & \omega^2 & \cdots & \omega^{2^n-1} \\ 1 & \omega^2 & \omega^4 & \cdots & \omega^{2(2^n-1)} \\ 1 & \omega^3 & \omega^6 & \cdots & \omega^{3(2^n-1)} \\ \vdots & \vdots & \vdots & \ddots & \vdots \\ 1 & \omega^{2^n-1} & \omega^{2(2^n-1)} & \cdots & \omega^{(2^n-1)(2^n-1)} \end{bmatrix} \tag{7.5}$$

where $\omega = e^{2\pi i/2^n}$.

How do we get this matrix? To make it easier to understand we'll identify the important part of the summation we need for the matrix representation (which we'll label M_{xy}):

$$\sum_{x=0}^{2^n-1} \sum_{y=0}^{2^n-1} \frac{\alpha_x e^{2\pi ixy/2^n}}{\sqrt{2^n}} |y\rangle = \frac{1}{\sqrt{2^n}} \sum_{x=0}^{2^n-1} \left(\sum_{y=0}^{2^n-1} M_{xy} \, \alpha_x \right) |y\rangle \tag{7.6}$$

where $M_{xy} = e^{2\pi ixy/2^n}$.

Now using the summations, here are a few values of x and y for M_{xy}:

$$\frac{1}{\sqrt{2^n}} \begin{bmatrix} e^{2\pi i \cdot 0 \cdot 0/2^n} = e^0 = 1 & e^{2\pi i \cdot 1 \cdot 0/2^n} = e^0 = 1 & \cdots \\ e^{2\pi i \cdot 0 \cdot 1/2^n} = e^0 = 1 & e^{2\pi i \cdot 1 \cdot 1/2^n} = e^{2\pi i/2^n} = \omega & \\ \vdots & & \ddots \end{bmatrix}. \tag{7.7}$$

Next we'll look at two examples using the matrix representation of the QFT.

Example   A simple one qubit QFT. Given:

$$|\Psi\rangle = \frac{2}{\sqrt{5}}|0\rangle + \frac{1}{\sqrt{5}}|0\rangle.$$

Find,

$$|\Psi'\rangle = QFT\,|\Psi\rangle.$$

We'll use the matrix representation, which is:

$$\frac{1}{\sqrt{2}}\begin{bmatrix} 1 & 1 \\ 1 & e^{i\pi} \end{bmatrix}.$$

This matrix is actually just an H gate (since $e^{i\pi} = -1$), so we get:

$$|\Psi'\rangle = \frac{1}{\sqrt{2}}\begin{bmatrix} 1 & 1 \\ 1 & e^{i\pi} \end{bmatrix}\begin{bmatrix} \frac{2}{\sqrt{5}} \\ \frac{1}{\sqrt{5}} \end{bmatrix}$$

$$= \begin{bmatrix} \frac{2}{\sqrt{10}} + \frac{1}{\sqrt{10}} \\ \frac{2}{\sqrt{10}} + \frac{e^{\pi i}}{\sqrt{10}} \end{bmatrix}$$

$$= \begin{bmatrix} \frac{3}{\sqrt{10}} \\ \frac{2}{\sqrt{10}} + \frac{-1}{\sqrt{10}} = \frac{1}{\sqrt{10}} \end{bmatrix}$$

$$= \frac{3}{\sqrt{10}}|0\rangle + \frac{1}{\sqrt{10}}|1\rangle.$$

Example A two qubit QFT. Given:

$$|\Psi\rangle = \frac{1}{\sqrt{2}}|00\rangle + \frac{1}{\sqrt{2}}|11\rangle.$$

Find,

$$|\Psi'\rangle = QFT|\Psi\rangle.$$

The matrix representation is:

$$\frac{1}{\sqrt{4}}\begin{bmatrix} 1 & 1 & 1 & 1 \\ 1 & e^{\pi i/4} & e^{\pi i2/4} & e^{\pi i3/4} \\ 1 & e^{\pi i2/4} & e^{\pi i4/4} & e^{\pi i6/4} \\ 1 & e^{\pi i3/4} & e^{\pi i6/4} & e^{\pi i9/4} \end{bmatrix}.$$

So $|\Psi'\rangle$ is:

$$|\Psi'\rangle = \frac{1}{\sqrt{4}}\begin{bmatrix} 1 & 1 & 1 & 1 \\ 1 & e^{\pi i/4} & e^{\pi i2/4} & e^{\pi i3/4} \\ 1 & e^{\pi i2/4} & e^{\pi i4/4} & e^{\pi i6/4} \\ 1 & e^{\pi i3/4} & e^{\pi i6/4} & e^{\pi i9/4} \end{bmatrix}\begin{bmatrix} \frac{1}{\sqrt{2}} \\ 0 \\ 0 \\ \frac{1}{\sqrt{2}} \end{bmatrix}$$

$$= \frac{1}{\sqrt{4}}\begin{bmatrix} \frac{1}{\sqrt{2}} + 0 + 0 + \frac{1}{\sqrt{2}} \\ \frac{1}{\sqrt{2}} + 0 + 0 + \frac{1}{\sqrt{2}}e^{\pi i3/4} \\ \frac{1}{\sqrt{2}} + 0 + 0 + \frac{1}{\sqrt{2}}e^{\pi i6/4} \\ \frac{1}{\sqrt{2}} + 0 + 0 + \frac{1}{\sqrt{2}}e^{\pi i9/4} \end{bmatrix}$$

$$= \frac{1}{\sqrt{4}}\begin{bmatrix} \frac{1}{\sqrt{2}} + 0 + 0 + \frac{1}{\sqrt{2}} \\ \frac{1}{\sqrt{2}} + 0 + 0 + \frac{-1+i}{\sqrt{2}} \\ \frac{1}{\sqrt{2}} + 0 + 0 + \frac{-i}{\sqrt{2}} \\ \frac{1}{\sqrt{2}} + 0 + 0 + \frac{1+i}{\sqrt{2}} \end{bmatrix}$$

$$= \begin{bmatrix} \frac{2}{\sqrt{8}} \\ \frac{\sqrt{2}-1+i}{4} \\ \frac{1-i}{\sqrt{8}} \\ \frac{\sqrt{2}+1+i}{4} \end{bmatrix}$$

$$= \frac{2}{\sqrt{8}}|00\rangle + \frac{\sqrt{2}-1+i}{4}|01\rangle + \frac{1-i}{\sqrt{8}}|10\rangle + \frac{\sqrt{2}+1+i}{4}|11\rangle.$$

7.3.1.1 *How Do we Implement a QFT?*

As stated before, a one qubit QFT has the following rather simple circuit:

A three qubit QFT has a more complicated circuit:

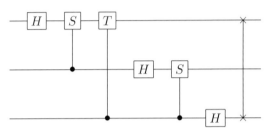

We can extend this logic to n qubits by using H and $n-2$ different rotation gates (R_k — see below) where:

$$R_k = \begin{bmatrix} 1 & 0 \\ 1 & e^{2\pi i/2^k} \end{bmatrix}. \tag{7.8}$$

For n qubits we use gates $R_2 \ldots R_n$. For more information on generic QFT circuits you'll need to consult an external reference like QCQI.

7.3.2 *Fast Factorisation*

Finding unknown factors p and q such that $p \times q = 347347$ is very slow compared to the reverse problem: calculating 129×156. We don't yet know if fast algorithms for factorising on classical machines are possible but we do now know (thanks to Shor) a fast factorisation algorithm we can run on a quantum computer.

Public key encryption systems like RSA algorithm rely on the fact that it is hard to factorise large numbers; if we could find the factors we could use the information provided in the public key to decrypt messages encrypted with it. In terms of RSA our task is simple. Given an integer N for which we know $n = pq$ where p and q are large prime numbers, we want to calculate p and q.

7.3.3 Order Finding

To factorise N we reduce the problem of factorisation to *order finding*, that is: given $1 < x < N$ the *order* of $x \bmod N$ is the smallest value of r where $r \geq 1$ and $x^r \bmod N = 1$.

This means that the list of powers of x, $1, x, x^2, x^3, x^4, \ldots \bmod N$ will repeat with a *period* that is less than N.

We're trying to find the period of a periodic function. Why? Because it turns out there's a close connection between finding the factors and finding the period of a periodic function. The intuition is that quantum computers might be good at this because of quantum parallelism: their ability to compute many function values in parallel and hence be able to get at "global properties" of the function (as in Deutsch's algorithm).

Example Say we have $N = 55$ and we choose x to be 13.

$$13^0 \quad \bmod 55 = 1,$$
$$13^1 \quad \bmod 55 = 13,$$
$$13^2 \quad \bmod 55 = 4,$$
$$\vdots$$
$$13^{20} \quad \bmod 55 = 1.$$

So $r = 20$, i.e. 13 mod 55 has a period of 20.

The calculation of $x^i \bmod N$ can be done in polynomial time and thus can be done on a classical computer. Once we have the order we can apply some further classical calculations to it to obtain a factor of N.

The "quantum" part gives us the period r in polynomial time by using a process called *phase estimation*. Phase estimation attempts to determine an unknown value γ of an eigenvalue $e^{2\pi i \gamma}$ of an eigenvector $|u\rangle$ for some unitary U. We won't worry about explicitly understanding phase estimation as Shor's algorithm has a number of specific steps that makes learning it unnecessary.

Before we look at the circuit for Shor's algorithm, a necessary component of Shor's algorithm, the continued fractions algorithm, is introduced.

7.3.3.1   *The Continued Fractions Algorithm*

The *continued fractions algorithm* allows us to calculate an array of integers to represent a fraction. We split the fraction into its whole and fractional parts, store the whole part, find the reciprocal of the fractional part and repeat the process until we have no fractional parts left. Here's an example:

Example   Convert $\frac{11}{9}$ to integer array representation.

$$\frac{11}{9} = 1 + \frac{2}{9}$$

$$= 1 + \cfrac{1}{\underset{\frac{9}{2} = 4\frac{1}{2}}{\frac{9}{2}}}$$

$$= 1 + \cfrac{1}{4 + \cfrac{1}{\underset{\frac{2}{1} = 2}{\frac{2}{1}}}}$$

So, we end up with the following list:

$$[1, 4, 2]$$

which is a three element array (that took three steps) that represents $\frac{11}{9}$.

Now we'll look at the fast factorisation algorithm...

7.3.3.2 *The Fast Factorisation Circuit and Algorithm*

The algorithm presented here (see figure 7.1) is similar to Shor's algorithm, and uses the QFT introduced earlier. It's simple — given a number N to be factored, return a factor f where $f > 1$. The algorithm is as follows:

(1) If N is divisible by 2 then return $f = 2$.
(2) For $a \geq 1$ and $b \geq 2$ if $N = a^b$ then return $f = a$ (we can test this classically).
(3) Randomly choose an integer x where $1 < x < N$. We can test if two numbers share a common divisor efficiently on a classical computer. There is an efficient classical algorithm to test if numbers are *coprime*, that is their greatest common divisor (gcd) is 1. In this step we test if $\gcd(x, N) > 1$, and if it is then we return $f = \gcd(x, N)$. E.g. if $N = 15$ and $x = 3$; we find $\gcd(3, 15) = 3$, so return 3.
(4) This is where the quantum computer comes in. We apply the quantum order finding algorithm. Before we start we need to define the size of the input registers. Register 1 needs to be t qubits in size where $2N \leq t$ (this is to reduce the chance of errors in the output). Register 2 needs to be L qubits in size where L is the number of qubits needed to store N.

$|\psi_1\rangle$ Initialise register 1, which is t qubits in size to $|0\rangle^{\otimes t}$ and register 2, which is L qubits in size to $|1\rangle^{\otimes L}$.

$|\psi_2\rangle$ Create a superposition on register 1:
$$|\Psi\rangle = \frac{1}{\sqrt{2^t}} \sum_{r_1=0}^{2^t - 1} |R_1\rangle |1\rangle.$$

$|\psi_3\rangle$ Apply $U_f R_2 = x^{R_1} \mod N$ to register 2:
$$|\Psi\rangle = \frac{1}{\sqrt{2^t}} \sum_{r_2=0}^{2^t - 1} |R_1\rangle |x^{R_1} \mod N\rangle.$$

$|\psi_4\rangle$ We measure register 2, because it is entangled with register 1 and our state becomes subset of the values in register 1 that correspond with the value we observed in register 2.

$|\psi_5\rangle$ We apply $QFT^\dagger$ to register 1 and then measure it.

$|\psi_6\rangle$ **(not shown)** Now we apply the continued fractions algorithm to $\frac{|\Psi\rangle}{2^t}$ and the number of steps it takes will be the period r.

(5) With the result r first test if r is even, check if $x^{r/2} \neq -1 \mod N$ then calculate $f = \gcd(x^{r/2} \pm, N)$. If the result is not 1 or N then return f as it is a factor, otherwise the algorithm has failed and we have to start again.

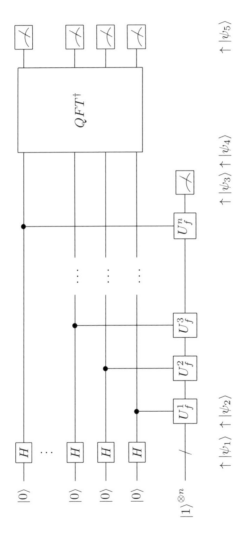

Fig. 7.1   The fast factorisation circuit

Over the next two pages we'll look at a couple of worked examples of Shor's algorithm.

Example Find the factors for $N = 15$

(1) N is not even, continue.

(2) $N \neq a^b$ so continue.

(3) We choose $x = 7$ $\gcd(7, 15) = 1$ so continue.

(4) $t = 11$ qubits for register 1 and $L = 4$ qubits for register 2.

$|\psi_1\rangle$ Initialise both the registers so the combined state becomes $|\Psi\rangle = |000000000001111\rangle$.

$|\psi_2\rangle$ Create a superposition on register 1, so we get $|\Psi\rangle = \frac{1}{\sqrt{2048}}(|0\rangle + |1\rangle + |2\rangle + \ldots + |2047\rangle)|15\rangle$.

$|\psi_3\rangle$ applying x^{R_1} mod 15 gives us the following:

R_1	$|0\rangle$	$|1\rangle$	$|2\rangle$	$|3\rangle$	$|4\rangle$	$|5\rangle$	$|6\rangle$	$|7\rangle$	$|8\rangle$	$|9\rangle$	$|10\rangle$	...
R_2	$|1\rangle$	$|7\rangle$	$|4\rangle$	$|13\rangle$	$|1\rangle$	$|7\rangle$	$|4\rangle$	$|13\rangle$	$|1\rangle$	$|7\rangle$	$|4\rangle$	...

$|\psi_4\rangle$ We measure R_2 and randomly get a 4 leaving $|\Psi\rangle$ in the following post measurement state:

R_1			$|2\rangle$				$|6\rangle$				$|10\rangle$	...
R_2			$|4\rangle$				$|4\rangle$				$|4\rangle$	...

Remember that both registers are actually part of the same state vector. It's just convenient to think of them separately. The state above is actually an entangled state that looks like (R_2 is **bolded**):

$$|\Psi\rangle = \frac{1}{\sqrt{512}}(|000000000010\mathbf{0100}\rangle + |000000000110\mathbf{0100}\rangle + \ldots).$$

$|\psi_5\rangle$ After applying $QFT^\dagger$ we get either $0, 512, 1024,$ or 1536 with a probability of $\frac{1}{4}$. Say we observe 1536.

$|\psi_6\rangle$ The result from the continued fractions algorithm for $\frac{1536}{2048}$ is 4.

(5) r is even and satisfies $\frac{r}{2} \neq -1$ mod N. So we try $\gcd(7^2 - 1, 15) = 3$ and $\gcd(7^2 + 1, 15) = 5$. Now, by testing that $3 \times 5 = 15 = N$ we see we have now found our factors.

Example Find the factors for $N = 55$

(1 & 2.) N is not even and $N \neq a^b$, so continue.

(3) We choose $x = 13$ gcd$(13, 55) = 1$ so continue.

(4) $t = 13$ qubits for register 1 and $L = 6$ qubits for register 2.

$|\psi_1\rangle$ Initialise both the registers so the combined state becomes $|\Psi\rangle = |0000000000000111111\rangle$.

$|\psi_2\rangle$ Create a superposition on register 1, so we get $|\Psi\rangle = \frac{1}{\sqrt{8192}}(|0\rangle + |1\rangle + |2\rangle + \ldots + |8191\rangle)|63\rangle$.

$|\psi_3\rangle$ applying $x^{R_1} \bmod 55$ gives us the following:

R_1	$|0\rangle$	$|1\rangle$	$|2\rangle$	$\ldots$	$|8192\rangle$
R_2	$|1\rangle$	$|13\rangle$	$|4\rangle$	$\ldots$	$|2\rangle$

$|\psi_4\rangle$ We measure R_2 and randomly get a 28 leaving $|\Psi\rangle$ in the following post measurement state:

R_1	$|9\rangle$	$|29\rangle$	$|49\rangle$	$\ldots$	$|8189\rangle$
R_2	$|28\rangle$	$|28\rangle$	$|28\rangle$	$\ldots$	$|28\rangle$

So the state vector (ie. with both registers) looks like this:

$$|\Psi\rangle = \frac{1}{\sqrt{410}}\left(|9\rangle|28\rangle + |29\rangle|28\rangle + |49\rangle|28\rangle + \ldots + |8189\rangle|28\rangle\right).$$

$|\psi_5\rangle$ After applying $QFT^\dagger$ we observe 4915 (the probability of observing this is 4.4%).

$|\psi_6\rangle$ The result from the continued fractions algorithm for $\frac{4915}{8192}$ is 20.

(5) r is even and satisfies $\frac{r}{2} \neq -1 \bmod N$. So we try gcd$(13^{10} - 1, 15) = 5$ and gcd$(13^{10} + 1, 55) = 11$. Now, by testing that $5 \times 11 = 55 = N$ we see we have now found our factors.

7.4 Grover's Algorithm

Grover's algorithm gives us a quadratic speed up to a wide variety of classical search algorithms. The most commonly given examples of search algorithms that can benefit from a quantum architecture are *shortest route finding* algorithms and algorithms to find specific elements in an unsorted database.

7.4.1 The Travelling Salesman Problem

An example of shortest route finding is the *traveling salesman problem*. Put simply, given a number of interconnected cities, with certain distances between them, is there a route of less than k kilometers (or miles) for which the salesman can visit every city?

With Grover's algorithm it is possible to complete a search for a route of less than k kilometers in $O(\sqrt{N})$ steps rather than an average of $\frac{N}{2}$ (which is $O(N)$) steps for the classical case. If we have M different solutions for k then the time complexity for Grover's algorithm is $O(\sqrt{\frac{M}{N}})$.

7.4.2 Quantum Searching

For the purpose of explaining Grover type algorithms we'll use an unsorted database table as an example.

Given a database table with N elements (it is best if we choose an N that is approximately 2^n where n is the number of qubits) with an index i, our table is shown below:

0	element 1
1	element 2
2	element 3
$\vdots$	
$N - 1$	element N

Suppose there are M solutions where $1 < M \leq N$.

In a similar way to Deutsch's algorithm we use an oracle to decide if a particular index, x is a *marked* solution to the problem, i.e.

$$f(x) = 1 \text{ if } x \text{ is a solution,}$$
$$f(x) = 0 \text{ otherwise.}$$

The search algorithm can actually be made up of several oracles. The oracle functions very similarly to the Deutsch Josza algorithm, as shown below:

$$|x\rangle |q\rangle \rightarrow O \rightarrow |q \oplus f(x)\rangle \tag{7.9}$$

here $|x\rangle$ is a register, $|q\rangle$ is a qubit and O is the oracle. The oracle circuit looks like this:

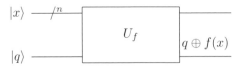

As with the Deutsch Josza algorithm if we set q to $|1\rangle$ and then put it through an H gate the answer appears in the $|x\rangle$ register while $|q\rangle$ appears the same after the calculation. So we end up with:

$$|x\rangle \to O \to (-1)^{f(x)} |x\rangle . \tag{7.10}$$

The function $f(x)$ contains the logic for the type of search we are doing. Typically there are extra work qubits leading into the oracle that may behave as ancilla qubits. This is called an oracle's *workspace* (represented by w).

The circuit for Grover's algorithm is shown below:

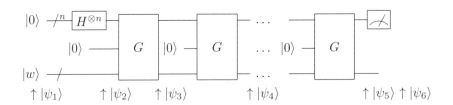

The steps for the algorithm for $M = 1$ are as follows:

$|\psi_2\rangle$ Initialise qubits states.
$|\psi_2\rangle$ We put $|x\rangle$ into a superposition:

$$|x\rangle \to \frac{1}{\sqrt{N}} \sum_{x=0}^{N-1} |x\rangle .$$

$|\psi_3\rangle \rightarrow |\psi_5\rangle$ Each G is called a *Grover iteration* that performs the oracle and a *conditional phase flip* on $|x\rangle$ which flips the sign on all qubits except for $|0\rangle$ (denoted by CPF below). This is done after collapsing the superposition on $|x\rangle$ via $H^{\otimes n}$. After the phase flip is completed the $|x\rangle$ register is put back into a superposition. Each G looks like the following:

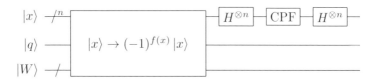

For $M = 1$ we will need to apply G $\lceil \pi\sqrt{2^n}/4 \rceil$ times.

$|\psi_6\rangle$ Finally we measure. As we have $M = 1$ the register $|x\rangle$ will contain the only solution. If we had $M > 1$ we would randomly measure one of the possible solutions.

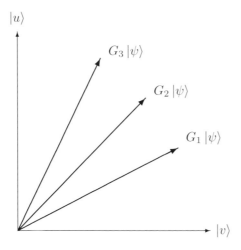

Fig. 7.2   Visualising grover's algorithm.

7.4.2.1   *Visualising Grover's Algorithm*

We can define the superposition of solutions to Grover's algorithm as:

$$|u\rangle = \frac{1}{\sqrt{M}} \sum_{=} |x\rangle \qquad (7.11)$$

and the superposition of values that are not solutions as:

$$|v\rangle = \frac{1}{\sqrt{N-M}} \sum_{\neq} |x\rangle . \qquad (7.12)$$

The operation of Grover's algorithm can be seen in figure 7.2 as a series of rotations of $|\Psi\rangle$ from $|v\rangle$ to $|u\rangle$. Each individual rotation is done by G, a Grover iteration.

A simple example follows:

Example   Suppose we have an index size of 4, which gives us $N = 4$. We have only one solution so $M = 1$ and our function, $f(x)$, has a marked solution at $x = 0$. This is like saying the element we are looking for is at index $i = 0$. These are the results returned by an oracle call:

$$f(0) = 1, f(1) = 0, f(2) = 0, \text{ and } f(3) = 0.$$

The size of register $|x\rangle$ is 2 qubits. We also have a workspace of 1 qubit (which we set to $|1\rangle$), which we put through an H gate at the same time as the qubits in $|x\rangle$ initially go through their respective H gates (we'll ignore oracle qubit q for this example). The steps for the algorithm are as follows:

$|\psi_1\rangle$ We initialise $|x\rangle$ and $|w\rangle$, so $|\Psi\rangle = |001\rangle$.

$|\psi_2\rangle$ $|x\rangle$ and $|w\rangle$ go through their H gates giving us $|x\rangle = \frac{1}{\sqrt{4}}[|00\rangle + |01\rangle + |10\rangle + |11\rangle]$ and $|w\rangle = \frac{1}{\sqrt{2}}[|0\rangle - |1\rangle]$.

$|\psi_3\rangle$ A single Grover iteration is all we need to rotate $|x\rangle$ to match $|u\rangle$ (the marked solution) so we jump straight to $|\psi_5\rangle$.

$|\psi_5\rangle$ Now $|x\rangle = |00\rangle$.

$|\psi_6\rangle$ Measuring register $|x\rangle$ gives us a 0.

Chapter 8

Using Quantum Mechanical Devices and Recent Developments

8.1 Introduction

Today's manufacturing methods can make chips (figure 8.1) with transistors a fraction of a micron wide. As they get smaller, chips also tend to get faster. According to Moore's law computer power doubles every few years and we'll soon reach the threshold below which transistors will not work because of quantum effects. At tens of nanometers (which we are close to now) electrons can tunnel between parts of a circuit, so transistor technology may not be able to shrink much further [4].

In 1998 Neil Gershenfeld (MIT) built a three qubit device, then in 2000 Ray LaFlemme built a seven qubit quantum computer. Currently we are still limited to tens of qubits and hundreds of gates with implementations being slow and temperamental. The architectures of these machines vary, with scientists even talking about using cups of coffee and small amounts of chloroform to build a working quantum computer [7].

As we have seen in previous chapters, classical computers still have a large part to play in quantum algorithms. For this reason it is likely that an architecture for a machine implementing a quantum algorithm would contain a classical computer (which acts as the master) and a quantum hardware interface to the quantum circuitry.

8.2 Physical Realisation

We'll take a quick look at physical realisations here, but a warning for those expecting detail — this section is mainly here for completeness and is strictly an introduction.

According to Nielsen and Chuang [31] there are four basic requirements

Fig. 8.1   A silicon chip.

for the physical implementation of a quantum computer. They are:

Qubit implementation — The biggest problem facing quantum computing is the fickle nature of the minute components they work with. Qubits can be implemented in various ways, like the spin states of a particle, ground and excited states of atoms, and photon polarisation. There are several considerations for implementing qubits. One consideration is high decoherence times (low stability), e.g. 10^{-3} seconds for an electron's spin, and 10^{-6} seconds for a quantum dot (a kind of artificial atom). Another consideration is speed. The stronger the qubit implementation can interact with the environment, the faster the computer. For example, nuclear spins give us much slower clock speed than electron spins, because of the nuclear spin's weak interactions with the outside world.

There are two types of qubits, material qubits like the stationary ones described above and *flying qubits* (usually photons). Stationary qubits are most likely to be used to build quantum hardware, whereas flying qubits are most likely to be used for communication.

Control of unitary evolution — How we control of the evolution of the circuits.

Initial state preparation (qubits) — Setting the values of initial qubits. We need to be able to initialise qubits to values like $|000\ldots\rangle$. This is not just for initial values of a circuit, e.g. quantum error correction needs a constant supply of pre-initialised, stable qubits.

Measurement of the final state(s) — Measuring qubits. We need a way of measuring the state of our qubits. We need to do this in a way that does not disturb other parts of our quantum computer. We also need to consider if the measurement is a *nondestructive* measurement, which for example leaves the qubit in the state which can be used later for initialisation. Another issue is that measurement techniques are less than perfect, so we have to consider copying values of the output qubits and averaging the results.

Also, David P. Divenco [18] has suggested two more fundamental requirements:

- Decoherence times need to be much longer than quantum gate times.
- A universal set of quantum gates.

8.2.1 *Implementation Technologies*

There are many theoretical ways to implement a quantum computer, all of which suffer from poor scalability at present. Two of the important methods are listed below [6].

Optical photon computer — This is the easiest type of quantum computer to understand. One of the ways qubits can be represented is by the familiar polarisation method. Gates can be represented by beamsplitters. Measurement is done by detecting individual photons and initial state preparation can be done by polarising photons. In practice, photons do not interact well with the environment. Although there are new methods that use entanglement to combat this problem there are still problems with single photon detection. Also, photons are hard to control as they move at the speed of light.

Nuclear Magnetic Resonance — (NMR) uses the spin of an atomic nucleus to represent a qubit. Chemical bonds between spins are manipulated by a magnetic field to simulate gates. Spins are prepared magnetically and induced voltages are used for measurement. Currently it is thought that NMR will not scale to more than about twenty qubits. Several atomic spins can be combined chemically in a molecule. Each element *resonates* at a different frequency so we can manipulate different spins by producing a radio wave pulse at the correct frequency. This spin is "rotated" by the radio pulse (the amount of which depends on the amplitude and direction). A computation is made up of a series of timed and sized radio pulses. We are not limited to using atoms as they can be combined to form a macroscopic liquid with same state spins for all the component atoms. A seven qubit computer has been made from five fluorine atoms whose spins implement qubits.

There are many more implementation technologies like ion traps (a number of ions trapped in a row in a magnetic field), SQUIDS (Superconducting Quantum Interference Devices), electrons on liquid helium, optical lattices, harmonic oscillators, cavity quantum electrodynamics (CQED), etc.

Finally, there is a quantum computing technology, Deterministic Quantum Computation with One pure qubit ($DQC1$), that is proposed to work mostly without entanglement.

The main quantum computing Wikipedia page has a good overview of current hardware technologies:
`http://en.wikipedia.org/wiki/Quantum_computer`.

8.3   Quantum Computer Languages

Even though quantum computing is in its infancy that hasn't stopped the proliferation of papers on various aspects of the subject. Many such papers have been written defining language specifications. There are a number of models of quantum computation including the quantum Turing machine, quantum circuit model, and more. We'll use a model called Quantum Random Access Machine (QRAM) [29]. QRAM is the most useful when discussing programming languages. The QRAM model contains classical and quantum registers and allows operations to be performed on both within in the same program. An architecture put forward in [39] consists of four layers:

High level programming language — Similar to classical high level languages but allows quantum operations.

Compiler — Compilation to quantum circuit machine code. Contains optimisations and can also be used to handle quantum error corrections.

Quantum Assembly language (QASM) — Maps directly to quantum circuit model instructions.

Quantum Physical Operations Language (QCPOL) — A language that maps quantum circuit model instructions to a particular physical implementation.

It is important to remember that all functions executed on a quantum computer must be reversible. In the case where an irreversible function needs to be executed on a quantum computer it can be converted into a reversible function by carrying the input with the result.

Some quantum languages are listed below [23, 29, 21].

QCL — (*Bernhard Ömer*) C like syntax and very complete. Accessible at http://tph.tuwien.ac.at/~oemer/qcl.html .

qGCL — (*Paolo Zuliani and others*) Resembles a functional programming language and claims to be better than Bernhard Ömer's QCL because QCL does not include probabilism and nondeterminism, has no notion of program refinement, and only allows standard observation. Accessible via http://web.comlab.ox.ac.uk/oucl/work/paolo.zuliani/ .

Quantum C — (*Stephen Blaha*) Currently just a specification, with a notion of quantum assembler. Accessible at http://arxiv.org/abs/quant-ph/0201082 .

Conventions for Quantum Pseudo Code — (*E. Knill*) Not actually a language, but a nice way to represent quantum algorithms and operations. Accessible at www.eskimo.com/~knill/cv/reprints/knill:qc1996e.ps .

LanQ — (*Mlnarik, H.*) Based on C and allows the creation of new processes and interprocess communication.

cQPL — This is a communication protocol aware version of QPL, a functional programming language. It is based on a QCL interpreter.

Q Language — (*Bettelli, S.*) Quantum computing class library for C++.

libquantum — Quantum computing class library for C.

λ^q**-calculus and** λ_q — (*Maymin, P.*) and (*van Tonder, A.*) Two versions of a quantum style λ-calculus.

QML — (*Altenkirch, T. and Grattage, J.*) Another functional programming language.

Question *It seems odd that there is no implementation of quantum BASIC. Is there any existing work? Maybe just a specification?*

8.4 Encryption Devices

The first encryption devices using the quantum properties discussed previously have been released. For example, a quantum key distribution unit developed by *id Quantique* (which can be found at `http://www.idquantique.com/`) is pictured in figure 8.2 and another encryption device has been released by *MagiQ*.

Fig. 8.2   id Quantique's QKD device.

8.5 Recent Developments

As of time of writing (2012) there has been many recent promising advances and the future looks bright for QC. Some recent hardware and architecture, cryptography and algorithmic developments are described next.

8.5.1 *Hardware and Architecture*

Great advances have been made in suppressing decoherence and reducing error rates. Other developments include: large scale entanglement (billions of atoms), quantum teleportation techniques have improved, and a substantial increase in qubit lifetimes. Architecturally a blueprint for quantum RAM has been developed, which fits in well with the development of the first *universally programmable quantum computer* and a qauntum version of Von Neumann architecture.

Commercial quantum computer company *D-Wave Systems* have made a number of impressive claims including the release of the world's first commercial quantum computer. These claims are yet to be verified.

8.5.2 *Cryptography*

Some early flawed QKD implementations have been successfully hacked (Like those sold by id Quantique). This does not mean that QKD is flawed in principle, just that the implementations have security holes. Another issue is denial of service, as the two points linked in a QKD system require an uninterrupted line of sight or a fibre optic cable. *Quantum networks* can be used to combat this issue.

Researchers have also looked at other quantum cryptographic schemes with varying levels of success. Examples of this include:

Quantum commitment — Alice commits to a value that she cannot change but can choose to reveal the value to Bob whenever she desires.

Position-based quantum cryptography — the geographic location of a participant determines whether or not the message is relayed successfully. E.g. Alice must be at a certain location to receive a message.

Finally, the field of *post-quantum cryptography* is emerging. This involves the creation of new classical encryption schemes that are secure against an attacker with a quantum computer.

8.5.3 *Algorithms*

While there have been many advances in the implementation of hardware there has been less advancement in the area of quantum algorithms. Any new class of algorithms would be of major importance. Some promising areas include such diverse areas as computer graphics, game theory, and graph traversal.

It has been shown that a *quantum random walk* (e.g. a random graph traversal) could traverse certain types of graph exponentially faster than any possible classical algorithm. Grover's algorithm can be viewed as a type of quantum random walk.

There's a good article on classical graphs and walks here:
http://en.wikipedia.org/wiki/Glossary_of_graph_theory.

Quantum game theory is the quantum analogue of *game theory* which deals with simple games like *the prisoner's dilemma* and is a surprising source of a number of papers relating to economics.

Here's an article on classical game theory: http://en.wikipedia.org/wiki/Game_theory.

Bibliography

[1] Baeyer, H. C. 2001, In the Beginning was the Bit, *New Scientist*, February 17

[2] Banerjee, S. 2004, *Quantum Computation and Information Theory — Lecture 1* [Online]. Available: `http://www.cse.iitd.ernet.in/\~suban/quantum/lectures/lecture1.pdf` [Accessed 14 May 2012]

[3] Barenco, A. Ekert, A. Sanpera, A. and Machiavello, C. 1996, *A Short Introduction to Quantum Computation* [Online]. Available: `http://www.Qubit.org/library/intros/comp/comp.html` [Accessed 30 June 2004]

[4] Benjamin, S. and Ekert, A. 2000, *A Short Introduction to Quantum-Scale Computing.* [Online]. Available: `www.dsc.ufcg.edu.br/~lula/cq-eliane/short.doc` [Accessed 14 May 2012]

[5] Bettelli, S. 2000, *Introduction to Quantum Algorithms* [Online]. Available: `sra.itc.it/people/serafini/quantum-computing/seminars/20001006-slides.ps` [Accessed 5 December 2004]

[6] Black, P.E. Kuhn, D.R. and Williams, C.J. 2000, *Quantum Computing and Communication* [Online]. Available: `http://hissa.nist.gov/~black/Papers/quantumCom.pdf` [Accessed 14 May 2012]

[7] Blume, H. 2000, *Reimagining the Cosmos.* [Online]. Available: `http://www.theatlantic.com/unbound/digicult/dc2000-05-03.htm` [Accessed 14 May 2012]

[8] Braunstein, S. L. and Lo, H. K. 2000, *Scalable Quantum Computers — Paving the Way to Realisation*, 1st edn, Wiley Press, Canada.

[9] Bulitko, V.V. 2002, *On Quantum Computing and AI (Notes for a Graduate Class).* [Online]. Available: `http://citeseerx.ist.psu.edu/viewdoc/summary?doi=10.1.1.112.7069` [Accessed 14 May 2012]

[10] Cabrera, B.J. 2000, *John von Neumann and von Neumann Architecture for Computers* [Online]. Available: `https://beacon.salemstate.edu/~tevans/VonNeuma.htm`[Accessed 14 May 2012]

[11] Castro, M. 1997, *Do I Invest in Quantum Communications Links For My Company?* [Online]. Available: `http://www.doc.ic.ac.uk/~nd/surprise_97/journal/vol1/mjc5/` [Accessed 14 May 2012]

[12] Copeland, J. 2000, *What is a Turing Machine?* [Online]. Available: `http:`

//www.alanturing.net/turing_archive/pages/reference%20articles/what%
20is%20a%20turing%20machine.html [Accessed 14 May 2012]

[13] cs.umbc.edu.edu 2003, *Complexity Class Brief Definitions* [Online]. Available: http://www.csee.umbc.edu/help/theory/classes.shtml [Accessed 14 May 2012]

[14] Dawar, A. 2004, *Quantum Computing — Lectures.* [Online]. Available: http://www.cl.cam.ac.uk/Teaching/current/QuantComp/ [Accessed 14 May 2012]

[15] Dorai, K. Arvind, Kumar, A. 2001, *Implementation of a Deutsch-like quantum algorithm utilising entanglement at the two-qubit level on an NMR quantum-information processor* [Online]. Available: http://eprints.iisc.ernet.in/archive/00000300/01/Deutsch.pdf [Accessed 14 May 2012]

[16] Designing Encodings 2000, *Designing Encodings* [Online]. Available: http://www.indigosim.com/tutorials/communication/t3s2.htm [Accessed 14 May 2012]

[17] Deutsch, D. and Ekert, A. 1998, Quantum Computation, *Physics World,* March

[18] Divincenzo, D.P. 2003, The Physical Implementation of Quantum Computation, *quant-ph/0002077,* vol. 13 April.

[19] Ekert, A. 1993, Quantum Keys for Keeping Secrets, *New Scientist,* Jan 16

[20] Forbes, S. Morton, M. Rae H. 1991, *Skills in Mathematics Volumes 1 and 2,* 2nd edn, Forbes, Morton, and Rae, Auckland.

[21] Gay, S. 2006 *Quantum Programming Languages Survey and Bibliography.*

[22] Gilleland, M. 2000, *Big Square Roots* [Online]. Available: http://www.merriampark.com/bigsqrt.htm [Accessed 14 May 2012]

[23] Glendinning, I. 2004, *Quantum Programming Languages and Tools.* [Online]. Available: http://www.vcpc.univie.ac.at/~ian/hotlist/qc/programming.shtml [Accessed 14 May 2012]

[24] Jones, J. Wilson, W. 1995, *An Incomplete Education*

[25] Knill, E. Laflamme, R. Barnum, H. Dalvit, D. Dziarmaga, J. Gubernatis, J. Gurvits, L. Ortiz, G. Viola, L. and Zurek, W.H. 2002, *Introduction to Quantum Information Processing*

[26] Marshall, J. 2001, *Theory of Computation* [Online]. Available: http://pages.pomona.edu/~jbm04747/courses/fall2001/cs10/lectures/Computation/Computation.html [Accessed 9 August 2004]

[27] McEvoy, J.P. and Zarate, 0. 2002, *Introducing Quantum Theory,* 2nd edn, Icon Books, UK.

[28] Meglicki, Z. 2002, *Quantum Complexity and Quantum Algorithms* [Online]. Available: http://beige.ucs.indiana.edu/B679/node27.html [Accessed 7 December 2004]

[29] Miszczak, J.A. 2011, *Models of Quantum Computation and Quantum Programming Languages*

[30] Nielsen, M. A. 2002, *Eight Introductory Lectures on Quantum Information Science* [Online]. Available: http://michaelnielsen.org/blog/introductory-lecture-notes-on-quantum-information-and-computation/ [Accessed 14 May 2012]

[31] Nielsen, M. A. and Chuang, I. L. 2000, *Quantum Computation and Quantum Information*, 3rd edn, Cambridge Press, UK.

[32] Rae, A. 1996, *Quantum Physics: Illusion or Reality?*, 2nd edn, Cambridge Press, UK.

[33] Shannon, C. E. 1948, *A Mathematical Theory of Communication* [Online]. Available: http://cm.bell-labs.com/cm/ms/what/shannonday/paper.html [Accessed 14 May 2012]

[34] Shatkay, H. 1995, *The Fourier Transform — A Primer* [Online]. Available: http://citeseer.ist.psu.edu/shatkay95fourier.html [Accessed 10 December 2004]

[35] Smithsonian NMAH 1999, *Jacquard's Punched Card* [Online]. Available: http://homepage.mac.com/oldtownman/recording/jacquard1.html [Accessed 14 May 2012]

[36] Stay. M. 2004, *Deutsch's algorithm with a pair of sunglasses and some mirrors* [Online]. Available: http://www.lepp.cornell.edu/spr/2004-04/msg0060395.html [Accessed 14 May 2012]

[37] Steane, A. M. 1998, Quantum Computing, *Rept.Prog.Phys.*, vol. 61 pp. 117-173

[38] [65] Svore, K.M. Cross, A.W. Aho, A.V. Chuang, I.L. and Markov, I.L. *Toward a software architecture for quantum computing design tools*, Proc. 2nd Int. Workshop on Quantum Programming Languages 1, CD-ROM (2004).

[39] [66] Svore, K.M. Cross, A.W. Aho, A.V. Chuang, I.L. and Markov, I.L. *A layered software architecture for quantum computing design tools*, Computer 39 (1), 7483 (2006).

[40] Wolfram, S. 2002, *A New Kind of Science*, 1st edn, Wolfram Media, USA

Image References

Figure 2.1
http://www.lindsredding.com/wp-content/uploads/2012/02/charles_babbage.
full_.jpg
http://upload.wikimedia.org/wikipedia/commons/2/2e/Ada_Lovelace_1838.jpg

Figure 2.2
http://www.eecs.berkeley.edu/~bh/ss-pics/alonzo.jpg
http://www.computerhistory.org/timeline/images/1954_turing_large.jpg

Figure 2.3
http://1.bp.blogspot.com/_5l4PuunDq3Y/S73UxVAISeI/AAAAAAAAAFA/p6DvAEa3cQc/
s1600/john-von-neumann.jpg

Figure 2.4
aima.cs.berkeley.edu/cover.html

Figure 2.5
http://www.phy.bg.ac.yu/web_projects/giants/hilbert.html
http://www-groups.dcs.st-and.ac.uk/~history/Mathematicians/Godel.html

Figure 2.9
http://www.creativitypost.com/images/uploads/technology/feynman-300.jpg

Figure 2.12
http://users.elis.ugent.be/~aldevos/projects/computer.html

Figure 3.9
http://4.bp.blogspot.com/--4hcGIDgxWI/TZ3CElFvQLI/AAAAAAAABrU/sPZGbWjSVV4/
s1600/Fourier2.jpg

Figure 4.2
http://images.suite101.com/1224831_com_514pxnikol.jpg
http://astro-canada.ca/_photos/a4303_galilee1_g.jpg

Figure 4.3
http://ww1.prweb.com/prfiles/2009/10/17/2122074/A8aDemocritus1.jpg
http://www.free-photos.biz/images/science/science_awards/john_dalton.jpg

Figure 4.4
http://esvc000605.wic057u.server-web.com/library/museum/images/Hermann%
20Helmholtz.jpg
http://www.solidariteetprogres.org/IMG/jpg/Rudolf_Clausius_01.jpg

Figure 4.6
http://www.uni-graz.at/uarc1www_boltzmannludwig.jpg
http://www.freeinfosociety.com/media/images/1186.jpg

Figure 4.7

http://www.faithinterface.com.au/wp-content/uploads/2011/03/
albert-einstein.jpg
http://en.wikipedia.org/wiki/File:Balmer.jpeg

Figure 4.8

http://adam.humanisti.sk/wp-content/2007/10/max_planck.jpg
http://images.travelpod.com/tripwow/photos2/ta-012d-371a-f3f2/
he-was-a-research-student-under-j-j-thomson-n5-cambridge-new-zealand+
13013580184-tpfil02aw-13352.jpg

Figure 4.9

http://ww1.prweb.com/prfiles/2009/10/12/2122074/A111bErnestRutherford2.jpg
http://www.scientific-web.com/en/Physics/Biographies/images/
ArnoldSommerfeld.jpg

Figure 4.13

http://osulibrary.oregonstate.edu/specialcollections/coll/pauling/catalogue/
09/1926i.61-pauli-600w.jpg
http://www.kof.zcu.cz/st/dp/horsky/html/broblie2.jpg

Figure 4.14

https://osulibrary.oregonstate.edu/specialcollections/coll/pauling/bond/
pictures/portrait-schrodinger-large.html
http://osulibrary.oregonstate.edu/specialcollections/coll/nonspcoll/
catalogue/portrait-heisenberg-900w.jpg

Figure 4.15

http://www.sil.si.edu/digitalcollections/hst/scientific-identity/fullsize/
SIL14-B5-09a.jpg
http://2.bp.blogspot.com/_z7-e_liVX3c/S__ySlOue6I/AAAAAAAABuM/65GteHmqQ1g/
s1600/Paul_Dirac_1902-1984.jpg

Figure 6.1

ttp://archive.computerhistory.org/resources/still-image/Chess_temporary/
still-images/2-1a.MIT_Bell_Laboratories.Shannon-Claude.BELL_LABORATIES.19XX.
L062302010..jpg
http://upload.wikimedia.org/wikipedia/commons/6/6b/Charles_Babbage_-_1860.jpg

Figure 8.1

http://www.qubit.org/oldsite/intros/nano/nano.html

Figure 4.1

http://publicdomainclip-art.blogspot.com.au/2010/05/james-clerk-maxwell.html
http://www.fromoldbooks.org/Aubrey-HistoryOfEngland-Vol3/pages/vol3-401-Sir-
Isaac-Newton/vol3-401-Sir-Isaac-Newton-q75-1002x1035.jpg

Figure 8.2

http://www.idquantique.com/

Index

λ^q-calculus, 229
λ_q, 229
$\frac{\pi}{8}$ gate, 137

Absorbtion spectrum, 101
Adjoint, 72
Alhazen, 94
Analytical engine, 4
Ancilla bits, 26, 148
AND gate, 14
Angles in other quadrants, 34
Anti-commutator, 81
Aristotle, 95
Atoms, 95
Automaton, 8

Babbage, Charles, 4
Balmer, Johann Jakob, 101
Basis, 57
BB84, 199
Bell state circuit, 150
Bell states, 132, 188
Bennet, C.H., 196
Bennett, Charles, 26
Big O notation, 19
Binary entropy, 171
Binary numbers, 7
Binary representation, 7
Binary symmetric channels, 180
Bit swap circuit, 148
Bits, 8
Black body, 99

Black body radiation, 99
Bloch sphere, 122, 172
Bohr, Niels, 98
Boltzmann's constant, 98
Boltzmann, Ludwig, 97
Boole, George, 164
Boolean algebra, 164
Born, Max, 107
Bra, 55
Bra-ket notation, 62
Brassard, G., 196
Bright line spectra, 101
Bubble sort, 18

Caesar cipher, 195
Cauchy–Schwartz inequality, 62
Cause and effect, 95
Cavity, 99
Channel capacity, 165
Characteristic equation, 74
Characteristic polynomial, 74
Christmas pudding model, 103
Church, Alonzo, 6
Church–Turing thesis, 6
Classical circuits, 13
Classical cryptography, 195
Classical error correction, 180
Classical gates, 13
Classical information sources, 166
Classical physics, 94
Classical redundancy and
    compression, 168

Classical registers, 13
Classical wires, 13
Clausius, Rudolf , 96
CNOT gate, 26, 144
Code wheel, 195
Codewords, 180
Column notation, 53
Commutative law of multiplication, 107
Commutator, 81
Compiler, 228
Completely mixed state, 176
Completeness relation, 72
Complex conjugate, 40
Complex number, 37
Complex plane, 41
Complex vector space, 53
Computational basis, 57
Computational resources and efficiency, 17
Constant coefficients, 32
Continued fractions algorithm, 216
Continuous spectrum, 101
Control lines, 26
Control of unitary evolution, 226
Controlled U gate, 148
Conventions for quantum pseudo code, 229
Converting between degrees and radians, 33
Copenhagen interpretation, 109
Copernicus, Nicolaus, 94
Copying circuit, 149
cQPL, 229
CROSSOVER, 16
Cryptology, 195
CSS codes, 187

D-Wave Systems, 230
de Broglie, Louis, 106
Decision problem, 21
Decoherence, 172, 181
Degenerate, 74
Degrees, 33
Democritus, 95
Density matrix, 173

Determinant, 49
Determinism, 95
Deterministic Turing machines, 21
Deutsch's algorithm, 202
Deutsch, David, 23
Deutsch–Church–Turing principle, 23
Deutsch–Josza algorithm, 207
Diagonal polarisation, 118
Diagonalisable matrix, 80
Dirac, Paul, 108
Discrete fourier transform, 85
Dot product, 52, 60
DQC1, 228
Dual vector, 55

Eigenspace, 74
Eigenvalue, 74
Eigenvector, 74
Einstein, Albert, 100
Electromagnetism, 94
Electron, 102
Emission spectrum, 101
Ensemble of states, 172
Ensemble point of view, 173
Entangled states, 131
Entanglement, 113
Entropy, 96
EPR, 113
EPR pair, 132, 188
Error syndrome, 184
Excited state, 104
Expectation value, 58
Exponential form, 43

FANIN, 16
FANOUT, 16
Fast factorisation, 214
Fast factorisation algorithm, 217
Fast factorisation circuit, 218
Feynman, Richard, 22
Finite state automata, 10
First law of thermodynamics, 96
Fluctuation, 98
Flying qubits, 226
For all, 32
Formal languages, 8

Four postulates of quantum
   mechanics, 155, 177
Fourier series, 86
Fourier transform, 85
Fourier, Jean Baptiste Joseph, 85
Fredkin gate, 28, 146
Frequency domain, 85
Full rank, 49
Fundamental point of view, 177

Gödel's incompleteness theorem, 5
Gödel, Kurt, 5
Galilei, Galileo, 94
Garbage bits, 26, 148
Global phase, 121
Global properties of functions, 202
Gram Schmidt method, 67
Ground state, 103
Grover workspace, 222
Grover's algorithm, 220, 231
Grover, Lov, 115

Hadamard gate, 137
Halting problem, 11
Heisenberg uncertainty principle, 109
Heisenberg, Werner, 107
Helmholtz, Von, 96
Hermitian operator, 77, 80
High level programming language,
   228
Hilbert space, 50, 62, 155
Hilbert, David, 5

ID Quantique, 230
Identity matrix, 48
If and only if, 32
Imaginary, 37
Imaginary axis, 41
Independent and identically
   distributed, 166
Initial point, 50
Inner product, 52, 60
Interference, 110
Intractable, 21, 201
Inverse matrix, 48

Inverse quantum fourier transform,
   210
Invertible, 24

Jacquard, Joseph Marie, 4
Joule, 98

Kelvin, 98
Ket, 53, 122
Keyes, R.W., 26
King, Ada Augusta, 4
Kronecker product, 84

Landauer's principle, 24
Landauer, Rolf, 24
LanQ, 229
Least significant digit, 8
Length of codes, 169
libquantum, 229
Linear combination, 56
Linear operator, 67
Linearly independent, 57
Local state, 113
Logarithms, 37
Logical symbols, 32

MagiQ, 230
Marked solution, 221
Markov process, 156
Matrices, 45
Matrix addition, 46
Matrix arithmetic, 46
Matrix entries, 46
Matrix multiplication, 47
Maxwell's demon, 25
Maxwell, James Clerk, 94
Measurement of final states, 227
Message destination, 165
Message receiver, 165
Message source, 164
Message transmitter, 164
Mixed states, 171
Mixture of states, 172
Modulus, 39
Moore's law, 5
Moore, Gordon, 5

Multi qubit gates, 144
Mutual key generation, 199

NAND gate, 15
Neumann, Jon Von, 5
New quantum theory, 98
Newton, Issac, 94
Newtonian mechanics, 94
No programming theorem, 154
Noisy channels, 179
Nondestructive measurement, 227
Nondeterministic polynomial time, 21
Nondeterministic Turing machines, 21
NOR gate, 14
Norm, 39
Normal operator, 77
Normalise, 64
NOT gate, 13
NOT$_2$ gate, 145
NP, 21
Nuclear magnetic resonance, 227
Nuclear spins, 113

Observable, 131
Old quantum theory, 98
One-time PAD, 196
Optical photon computer, 227
OR gate, 14
Oracle, 202
Order, 215
Order finding, 215
Orthogonal, 63
Orthonormal basis, 65
Outer product, 68
Outer product notation, 140

P, 21
Partial measurement, 126
Pauli exclusion principle, 105
Pauli gates, 134, 141
Pauli operators, 78
Pauli, Wolfgang, 105
Period, 215
Phase estimation, 215
Phase gate, 136
Photoelectric effect, 100

Planck's constant, 100
Polar coordinates, 41
Polar decomposition, 82
Polar form, 39
Polarisation of photons, 113
Polynomial time, 20
Polynomially equivalent, 23
Polynomials, 32
Position-based quantum
     cryptography, 231
Positive operator, 77, 80
Post-quantum cryptography, 231
POVMs, 160
Principle of invariance, 17
Prisoner's dilemma, 232
Probabilistic Turing machine, 22
Probability amplitudes, 59
Probability distribution, 166
Programmable quantum computer,
     154
Projective measurements, 131
Projectors, 69, 131, 157
Public key encryption, 195
Pure states, 171
Pythagorean theorem, 33

Q language, 229
QASM, 228
QCL, 229
QCPOL, 229
qGCL, 229
QML, 229
QPL, 229
Quantised, 93
Quantum assembly language, 228
Quantum bits, 116
Quantum C, 229
Quantum circuits, 133
Quantum commitment, 231
Quantum computer languages, 228
Quantum cryptography, 196
Quantum dot, 226
Quantum fourier transform, 210
Quantum fourier transform circuits,
     214
Quantum game theory, 231

Quantum hardware interface, 225
Quantum information sources, 171
Quantum key distribution, 198
Quantum logic gates, 133
Quantum mechanics, 93
Quantum money, 196
Quantum networks, 231
Quantum noise, 181
Quantum numbers, 104
Quantum packet sniffing, 197
Quantum Physical Operations
    Language, 229
Quantum random access memory, 228
Quantum random walk, 231
Quantum register, 116
Quantum repetition code, 182
Quantum searching, 221
Quantum Turing machine, 23
Qubit implementation, 226
Qubit initial state preparation, 226
Qubits, 116
Qubyte, 125
Qudit, 126
Quick sort, 18

Radians, 33
Random walk, 231
Rank, 49
Rationalising and dividing complex
    numbers, 43
Rayleigh–Jeans law, 106
Rectilinear polarisation, 118
Reduced density matrix, 177
Relative phase, 120
Repetition, 179
Repetition codes, 180
Reversible circuits, 29
Reversible computation, 26
Reversible gates, 26
Reversibly, 24
Right angled triangles, 33
Rotation operators, 142
RSA, 195, 201, 214
Rutherford, Ernest, 103
Rydberg constant, 101

Scalar, 46
Scalar multiplication by a matrix, 46
Schrödinger's cat, 108
Schrödinger, Erwin, 107
Schumacher compression, 178
Schumacher's quantum noiseless
    coding theorem, 172
Second law of thermodynamics, 96
Shannon entropy, 169
Shannon's communication model, 164
Shannon's noiseless coding theorem,
    169
Shannon, Claude E., 163
Shor code, 187
Shor's algorithm, 210
Shor, Peter, 115
Shortest route finding, 220
Simon's algorithm, 201
Simultaneous diagonalisation
    theorem, 82
Single qubit gates, 133
Single value decomposition, 82
Singular, 49
Sommerfeld, Arnold, 105
Source of noise, 164
Spanning set, 57
Spectral decomposition, 82
Spin, 105
Square root of NOT gate, 137
Stabiliser codes, 187
State vector, 59, 117
Statistical correlation, 113
Statistical mechanics, 97
Steane code, 187
Strong Church–Turing thesis, 22
Subsystem point of view, 177
Superdense coding, 150
Superposition, 110

Teleportation circuit, 152
Tensor product, 83, 126
Terminal point, 50
There exists, 32
Thermodynamics, 96
Thomson, Joseph J., 102
Time domain, 85

Toffoli gate, 27, 146
Trace, 75
Transpose matrix, 49
Trap door function, 196
Traveling salesman problem, 221
Trigonometric inverses, 34
Trigonometry, 33
Trigonometry identities, 36
Truth tables, 13
Turing machine, 6
Turing, Alan, 4

Uncertainty, 113
Uncomputable, 6
Unit vector, 64
Unitary operator, 77
Universal computer, 4
Universal Turning machine, 10
Universally programmable quantum
    computer, 230

Vector addition, 55
Vector scalar multiplication and
    addition, 54
Vectors, 50
Visualising grover's algorithm, 224
Von Neumann architecture, 5, 230
Von Neumann entropy, 173, 178

Wien's law, 106
Wiesner, Stephen, 196
Wire, 133
Witnesses, 21

XOR gate, 15

Zero matrix, 48
Zero memory information sources,
    167
Zero vector, 54

Wise, Mr., 8.

wish of the sisters, 61.

Wood, James P., Bishop of Philadelphia, 103-104.

woods, 21, 25, 52, 88, 155.

work, *vi, xxiii,* 2, 5, 14, 18, 45, 47, 57-58, 60-62, 70, 80, 87, 89, 93, 102, 104, 117, 124, 134, 136, 140, 142, 157, 160, 175, 176.

wounded, *v-vii, ix, xi-xiv, xvii-xix, xxi-xxiii,* 2-4, 8, 12-13, 15, 17-20, 32-33, 37-39, 46-47, 49-50, 53-54, 60, 77, 81, 85, 87-94, 98, 100-102, 109, 112-113, 119, 123, 130-131, 133, 140, 143, 145, 148, 150-153, 157-160, 163, 167, 169, 172-173, 175.

Yankees, 30.